Big Studio Secrets for Home Recording and Production

Joe Dochtermann

Course Technology PTR
A part of Cengage Learning

COURSE TECHNOLOGY
CENGAGE Learning™

Australia • Brazil • Japan • Korea • Mexico • Singapore • Spain • United Kingdom • United States

COURSE TECHNOLOGY
CENGAGE Learning™

Big Studio Secrets for Home Recording and Production

Joe Dochtermann

Publisher and General Manager,
Course Technology PTR:
Stacy L. Hiquet

Associate Director of Marketing:
Sarah Panella

Manager of Editorial Services:
Heather Talbot

Marketing Manager: Mark Hughes

Executive Editor: Mark Garvey

Project Editor: Kezia Endsley

Technical Reviewer:
Ashley Shepherd

Copy Editor: Kezia Endsley

Compositor: MPS Limited,
A Macmillan Company

Cover Designer: Luke Fletcher

Proofreader: Mike Beady

Indexer: Kelly Talbot

CD-ROM Producer:
Brandon Penticuff

For product information and technology assistance, contact us at
Cengage Learning Customer & Sales Support, 1-800-354-9706

For permission to use material from this text or product, submit all requests online at **cengage.com/permissions**

Further permissions questions can be emailed to
permissionrequest@cengage.com

Library of Congress Control Number: 2009942405

ISBN-13: 978-1-4354-5505-4
ISBN-10: 1-4354-5505-3

Course Technology, a part of Cengage Learning
20 Channel Center Street
Boston, MA 02210
USA

Cengage Learning is a leading provider of customized learning solutions with office locations around the globe, including Singapore, the United Kingdom, Australia, Mexico, Brazil, and Japan. Locate your local office at: **international.cengage.com/region**

Cengage Learning products are represented in Canada by Nelson Education, Ltd.

For your lifelong learning solutions, visit **courseptr.com**

Visit our corporate website at **cengage.com**

Printed in the United States of America
1 2 3 4 5 6 7 12 11 10

"If it sounds good, it *is* good."
—*Duke Ellington*

Acknowledgments

Thanks to the artists who have guided and inspired me to play guitar, record, and produce, and thereby enjoy music and life all the more:

To Charlie Karp, an amazing guitarist and singer, for his inspiration, and incredible times in the studio, on the stage, and all over creation.

To James Prosek for being my musical foil, a great friend, and a fellow aficionado of wildlife and the outdoors, and to Rick Richter, our partner in Troutband, for putting up with both of us!

To Dennis Hrbek, a friend, fool, hero, and fine engineer, for all our conversations about music, and for our fishing marathons in the summer.

To Harvey Brooks for his kindness, patience, and for teaching me how to play a mean bass guitar.

To Danny Kortchmar for his insights into producing music, and for letting me get away with stealing some guitar licks!

Thanks also to Mark Garvey at Cengage Learning for approaching me with this project, and for his guidance and encouragement, and to Kezia Endsley and Ashley Shepherd for their astute and good-humored editing work.

The book and CD-ROM have been written, recorded, and edited during our move to Berlin, and my wife Michaela and our wonderful friends in Berlin deserve special thanks for their support during this project. You mean the world to me as my best friends and creative co-conspirators in all we do. Thank you!

About the Author

Joe Dochtermann became interested in recording at age 16, after receiving a Tascam 4-track as a present. While playing guitar and recording with his high school band, it didn't take long for him to push that little 4-track tape recorder to its limits. Within a few years, he had assembled a 24-track studio, and began studying Music and Sound Recording at the University of New Haven in Connecticut. His passion for recording and production soon landed him work with producer Danny Kortchmar, singer Charlie Karp, bass guitar legend Harvey Brooks, and many other music professionals in the greater New York area to whom he owes his real-world music education.

After upgrading his recording rig to a professional Pro Tools setup, Joe honed his engineering and production chops, and worked extensively in New York, South Carolina, and in the Bahamas for clients including MCA records, V2 records, acclaimed classical violinist Alexander Markov, and blues-meisters SloLeak. Joe's latest project with his band Troutband brought him and bandmates James Prosek and Rick Richter to the late Johnny Cash's private studio to record with Nashville legends Marcus Hummon, David Roe, Rick Lonow, and Chuck Turner.

Joe now lives in Berlin, Germany. He is an avid guitar player, and since moving to Europe, he has written for *Guitar* magazine in Germany and published a series of music instruction books, including *The Charlie Christian Method for Jazz Guitar Improvisation, Master the Fretboard: Applied Theory for Guitarists, The Chord and Harmony Guide for Guitar,* and *Pro Techniques for Home Recording.* His books are available on Amazon, eBay, and via the author's website www.joedocmusic.com. Joe also continues to compose and produce music with his band project, Troutband, which can be found at www.troutband.com.

Contents

Chapter 4
Setting Up Your Studio   57

Chapter 5
Microphone Technique   83

Chapter 6
Signal Processing Toys: EQ   127

Chapter 7
Signal Processing Toys: Compression   151

Introduction

When music polymath Les Paul began experimenting with multi-track recording in his garage studio in 1947, he made a splash that would swell into a music technology tsunami in the following decades. As multi-track tape decks became available, a new art form was born—the art of studio recording.

Producers, artists, and engineers began making their fortunes by using technology to manipulate sound. Sam Phillips, Les Paul, Joe Meek, The Beach Boys, George Martin, Phil Spector, and Frank Zappa; the greatest works of these producers/artists would not have been possible without the manipulations of sound made possible by creative experimentation with technology.

In many cases, the recordings made by these pioneers used technology that would be considered laughable by today's standards. But does this mean that anyone with a lightning-fast PC, a spanking new audio interface, and a handful of microphones is guaranteed to make recordings that are just as good as these artists? The answer is a rather obvious no.

It seems to be the obvious answer, and so it begs the obvious question: Why?

If the technology is here, what is stopping us? Is it that digital technology lacks a certain warmth or soul? Is it a general lack of talent in the music world?

Nothing is stopping us. Technology never had soul. And with more and more people around, there is likely to be more talent in the world today than there ever was before; it just may not be found by large corporations with their eyes on the bottom line.

Many potential producer/engineers just don't know where to begin the recording journey, or they wind up stuck in a rut along the way. Those who aren't technically inclined become overwhelmed by the glut of software and hardware. Those who are creative in one way, such as in songwriting, may drown in the endless offering of plug-ins, samples, loops, effects libraries, and sound palettes carried in on the crest of this technological tidal wave.

Feeling the heavy expectations laid on you as a modern multi-tasker, you must be engineer, producer, arranger, performer, mixing and mastering engineer all rolled into one. This is a surefire recipe for frustration and disappointment.

This book explores the hows and whys of recording music in a modern context, and helps you determine how to use the intense technology available to create sounds that engage listeners—sounds that help define your identity as a producer, engineer, and/or performer.

First of all, you must know the rules in order to break them. Because recording is at least a partially technical art, ignorance is not bliss. These rules include the basics of acoustics, microphone technique, studio signal flow, and sound processing.

Next, the advancing producer/engineer must understand how to make technology work in order to cast off the technological shackles. Imagine Les Paul working in his garage, creating the first multi-layered recordings. He was using technology as a means to an end, to capture the sounds he imagined. What now seems like an endless expanse of possibilities in programs and plug-ins is often just a well-defined prison for creativity. Simulators, processors, sound libraries, and tone modelers are not a means to an end, but the illusion of an end in itself. This book looks at ways to make these things work for you by twisting their little heads and blending them with some classic studio techniques.

Finally, you need to consider the expectations and limitations of delivering the music. Mixing and mastering provide even the seasoned pro with new challenges in every song. You need to pull the music back out of that laptop and get it to where it was intended all along—to the listeners!

This book is a guide to studio technique—a potent package of tips, tricks, and techniques—but it is equally intended to keep you questioning at every turn, so don't hesitate to (metaphorically speaking) stop along the path to listen to the brook, smell a flower, or to turn over a stone and see what's hiding underneath.

Who Should Read This Book

This book is meant for anyone with an interest in both the artistic and technical sides of music and sound recording. The material is presented in an easy language, preferring the practical to the technical, teaching by using anecdotes and practical examples. Sidebars on technical topics allow you to delve deeper into the technical and theoretical side of recording, thereby keeping the main text easy and fun to read.

It is a perfect resource for anyone who feels daunted by the gear and software involved in audio recording, and overwhelmed by the flood of information (often marketing-driven, amateurish, or inaccurate) about recording presented on the web. The professional experience of the author and editors of this book provide solid information on studio recording and production, which is not biased by manufacturers' marketing ploys or bloggers' hot-headed opinions!

The concepts discussed in the book are backed up with real-world examples on the CD-ROM, presenting you with techniques that can be used in a *home studio* environment, although these are techniques regularly used in "big studios" across the world. The focus remains on the home studio throughout, and suggestions for improving your recordings are not always determined by the size of your wallet; even the simplest of home studios will benefit greatly from the tips, tricks, and techniques presented within.

What to Expect by Reading This Book

If you want to know how to use your recording gear to the best of its potential, no matter how humble or complex your studio, this book teaches you the technical prowess needed to make solid recordings, as well as a range of tips and tricks to make your recordings stand out in the crowd.

You can expect to glean ideas from classic recordings, understand modern techniques, and—with a little hard work—create the sounds that might find your work a place in the next generation of influential music. Many famous recordings, ones that sound excellent even by current standards, were made with simpler equipment than most modern home studios have available. You'll learn how to improve your sound from the bottom up. You'll learn about microphones and microphone technique, how acoustics affect recordings (and how to improve your studio on a shoestring budget), how to EQ, compress, and use effects on your tracks, how to produce, arrange, and mix, and how to master your recordings.

What's on the CD-ROM

The accompanying CD-ROM guides you through each chapter in a practical way. While the book provides you with the information behind music recording, the CD-ROM il- lustrates the examples with graphics and audio, providing a practical way of learning all the concepts. Seeing and hearing the concepts in action provides you with a whole new dimension in learning about music recording.

The CD-ROM works in a familiar website-style interface, organized into chapters. Usually, books with CDs require you to flip through the pages while holding one finger on the CD player and skipping through a string of examples; how annoying! This CD-ROM is a different concept altogether; it allows you to focus on the audio and visual explanations, with concise summaries of the text to accompany the examples. If you've ever used a mouse to surf the web, you're ready to go!

A Note from the Author

There is a particular fascination to recording sound. Like a child catching fireflies in a jam jar, you're capturing a moment hoping to enjoy it again and again. Recording sound can be that simple, but learning the technical side can be overwhelming. Of utmost importance is never to lose the feel and passion while tackling the technical side. When making music, you should always feel like you're running barefoot through the grass chasing those fireflies.

Working in a professional studio can be a great experience, and can do your music a lot of good. But the professional prices and the pressure of the ticking clock can be stifling. For this reason, many newbies turn to the relaxed atmosphere of the garage, spare bedroom, or semi-finished basement to create music recordings. What I do throughout this book is extract elements of the big professional studio and help you infuse them into your home environment.

Attention to details, preparation, experimentation, and the desire to develop a critical ear will help you make recordings at home that are as good as those coming out of professional studios. This book will help you develop those skills, with tips, tricks, and techniques that come from my experience working on projects done in a blend of big studio and home environments, often with experienced professionals.

In many cases, your microphones, audio interfaces, and DAW programs are the same as those used by the big studios. Your music is just as viable as any major artist's music. All that is missing is the technique and experience that has collected in these professional studio environments, which many folks never get a chance to experience.

Aside from learning about the workings of a recording studio from the ground up, you'll also learn about innovations in recording and how they came about. In contrast to many technical manuals, which tell you how things *must* be done, I focus on posing questions to you, in order to find what it is that you can bring to the recording process to make it unique. Music recordings sound good when properly engineered, but first sound *great* when everyone involved brings some passion into the work.

With that in mind, my goal is to make the book fun and easy to understand. I use anecdotes, illustrations, and practical examples. When you feel you have grasped some of the techniques in this book, go and apply them! You will need to try these techniques, listen, and adjust to suit your needs. Learning at its best is experimentation, and the wisdom that "creativity is 1% inspiration and 99% perspiration" most certainly applies.

References

Master Handbook of Acoustics (by F. Alton Everest), McGraw-Hill/TAB Electronics; Fourth edition (September 22, 2000)

How to Build A Small Budget Recording Studio From Scratch: With 12 Tested Designs (by Michael Shea), McGraw-Hill/TAB Electronics; Third edition (May 29, 2002)

Multi-track Recording for Musicians (by Brent Hurtig), Alfred Publishing Co., Inc. (October 1988)

Mastering Audio: The Art and the Science (by Bob Katz), Focal Press (2002). Additional references at
   Mr. Katz's website at www.digido.com

Pro Techniques for Home Recording (by Joe Dochtermann), Published through www.joedocmusic.com
   (September 2006)

Modern Recording Techniques (Huber & Runstein), Focal Press; Sixth edition (Feb 2, 2005)

The Beatles Recording Sessions, Mark Lewisohn (1988)

CD-ROM Downloads

If you purchased an ebook version of this book, and the book had a companion CD-ROM, we will mail you
a copy of the disc. Please send ptrsupplements@cengage.com the title of the book, the ISBN, your name,
address, and phone number. Thank you.

1 Looking Back for New Inspiration

If there's one thing for sure in the realm of music, it's that no one makes music on a deserted island; that is, all musicians get their inspiration from somewhere. Music is a social, communicative activity. Even if you choose to compose and record in your own private studio, you have heard music before, it has influenced you, and the music you are making is most likely meant to be heard by others.

Einstein once said "The secret to creativity is knowing how to hide your sources." This happens to be a standard *modus operandi* in the music field, especially when it comes to rock, country, and other forms of popular music, which have derived so much from traditional blues music as to often cross the line into outright theft. The trend continues, and hip-hop, electronic, and modern rock artists continue to act as "clever thieves," by quoting, sampling, and borrowing ideas both from their peers and from the tremendous catalog of recorded music stretching back into the 1920s.

Let's now take a brief look at a few of the most influential recording artists of the last half-century, and see what inspiration we might be able to glean from their concepts in terms of engineering and production. I am not talking about plagiarism here, but rather "standing on the shoulders of giants," to quote Sir Isaac Newton. This is where we stand now—armed with multi-channel audio interfaces, digital audio workstations (Pro Tools, Cubase, and so on), endless plug-ins, processors, and simulators, and an arsenal of affordable microphones. Up on the shoulders of all those who experimented and innovated before us, our assignment is to "do something fantastic" (as Frank Zappa would say) with all this stuff.

These innovators had their heads in the clouds, looking and listening for that next amazing sound, trying to take listeners to another level. We need to keep this task in the front of our minds as we go—let's *do something fantastic.*

Before Multi-Tracking—Sam Phillips

On July 5th, 1954, when Elvis Presley walked into producer/engineer Sam Phillips's Sun Records recording studio for an audition, chance and circumstance came together to change recording history.

Producer Sam Phillips had spent the early 1950s recording anyone and anything he could in the Memphis, Tennessee area to keep his dream of promoting "black" music alive. Phillips was connected to the direct, honest sound of the blues through his own rural childhood. He felt that this sound, if uniquely recorded and presented, would ring true to people everywhere.

He rented a small room in Memphis and set up a simple recording studio enhanced by some clever tricks of his own device. The recording gear consisted of an RCA 76D six-channel mono console, Ampex 350 mono tape machines (there was no stereo then!), some RCA ribbon mics, and the now infamous Shure 55 dynamic "Elvis mic."

By the early 50s, Phillips had enjoyed moderate success recording local blues artists, who often moved on to larger record labels. So he began searching for the right artist to make a crossover, to bring the sound and feel of black music to white audiences. Elvis's career wasn't an instant success, but over the course of a year working together with Phillips, they finally stumbled onto the arrangement of "That's All Right, Mama," which became Elvis's first hit. Even after selling Elvis's recording contract to RCA, Phillips went on to record Carl Perkins, Johnny Cash, Jerry Lee Lewis, and other artists who defined early rock n' roll.

Phillips felt that the music should be recorded clearly and simply, but with a certain twist, a "perfect imperfection," to make it stand out on the radio and in jukeboxes. As an engineer and producer, Phillips was always sure to highlight—or if necessary, create—a defining element within the music. Phillips's mixes placed Elvis's vocal in with the band, rather than loudly featured, as was more common at the time. When played on a juke-box, listeners heard not just a crooning voice backed by music, but a whole band that jumped out and danced!

This is a concept to which you can always return and take inspiration.

As a music genre or style develops a consistent sound, beat, arrangement, and so on, this is the time to create something that stands out from the crowd and grabs listener's attention again. As an engineer/producer, this is your task—to serve the music first and foremost while being prepared to search for something unique for the track. This could be a standout guitar tone, a unique vocal effect, an unusual arrangement that catches the listener off-guard, or just clearly capturing an artist's intense performance in all its glory.

By taking advantage of the studio's limitations in a creative way, Phillips accentuated the situation at hand, rather than trying to hide its potential flaws. His studio was a rather small 18×30 feet room, with 8-foot ceilings and no baffles to acoustically separate instruments. The walls were treated with radio-studio asbestos tiles, which tended to reflect midrange frequencies, creating a fairly "live" room acoustic. By using only a few microphones, Phillips recorded enough of the separate instruments to create a balanced recording, while avoiding the muddy sound caused by too many microphones in a live setting.

In Sun Studios, the mixer fed *two* Ampex 350 mono tape machines, using one to record and one to create an effect. This was Phillips's best-guarded and most often imitated studio trick (although Les Paul likely discovered this effect earlier, the popularity of Phillips's recordings put it on the map). The first tape machine was simply used as an echo generator, and the second recorded the performance. Here's how it went:

- The mics were set up as Phillips found fitting, generally using three or four mics, listening carefully, and sometimes moving performers closer to or farther away from mics to balance the sound.

- To use the first machine as an effect, Phillips sent the audio from the channels he wanted to have a slap-echo effect on to the recorder. The recording head printed the sound onto the running tape, which moved along quickly to the playback head, about an inch and a half down from the recording head.

- The sound from the *playback* head was then fed back into the mixing console. The time it took the tape to move from the recording head to the playback head was about 115 milliseconds (ms). When this was blended back in with the original signal, a slap-echo effect occurred. The song was mixed live and printed to the second tape machine.

Unique to this manner of creating a slap-echo is its tonal quality. Tape machines have certain inherent faults; they speed up and slow down slightly and the recording process somewhat dulls and compresses the sound. This particular tonal character has become the de facto standard for up-tempo rock n' roll, rockabilly, and country music echo effects.

INSIDE THE BOXES
Classic Rock n' Roll Slap-Echo   Try creating a simulation of this effect yourself. Create an auxiliary send and add a delay plug-in (or outboard delay) set to 115 milliseconds. In some cases you may want to time delays to the song (see Chapter 8, "Signal Processing Toys: Reverb and Delay-Based Effects"), but for this classic effect, you should try 115ms. See if you can modulate the delay time back and forth by a few percent—some

plug-ins have settings for this effect. Add a compressor and an EQ after the output of the delay. Compress the signal a bit, and roll off some of the low and high end with the EQ. Blend this channel back into your mix, and you can send any instrument you choose into rockabilly limbo at the twist of an Aux Send knob. It works particularly well on guitars (acoustic or electric), vocals, and even on percussive sounds, lending them a shaking, nervous energy that rubs against the main beat.

As a producer, Phillips was also unique. He worked with the artist in the search for what he called the "perfectly imperfect" performance. His idea was to capture a spontaneous energy that would fascinate listeners, not a technically perfect execution of the song.

The result was recordings that had an infectious energy, sounded uniquely of the Sun Studio room, and were engineered to sound great on radios and jukeboxes, the popular listening media of the day. This served both the interest of the listeners (a new and exciting sound) and Phillips's career (his recordings had a sound that identified him and his studio)—a brilliant concept. This is a critical point for you to keep in mind now, considering how many plug-ins and editing options are available to fix up what too many people consider to be mistakes. On those classic recordings, there is the occasional wrong note, off-mic vocal line, or rhythmic fluctuation. This is what makes these performances fascinating—these little "mistakes" keep the music sounding human and believable, and the focus is on the energy of performance, not on technical perfection.

Food for thought:

- Many current artists also leave in their "mistakes," even if they could be processed or edited out. Listen critically to some of your favorite music—there are likely certain imperfections, which you may have assumed were planned. When I love a song, in my mind it is perfect; it was meant to be recorded exactly that way. However, happy accidents are a matter of course when recording, as are flaws in a heartfelt performance. Encourage artists to let feeling come first, and try to polish this as little as possible.

- Consider how to manipulate modern listening situations to your favor, such as the exaggerated in-ear stereo of iPods, the low-fi of webcasts (and laptop speakers), and the radically extended low end of car and home stereos.

- Headphones and earbuds have again become popular (last time around was the Sony Walkman phase in the 1980s), and place the stereo field right inside the listener's head. Radically split stereo recordings went out of fashion after the 70s, but could be an interesting way to catch a younger listener's attention again.

- And as far as extended low end in stereos goes, consider using the automation features in modern DAWs to bring those high-energy basslines and drum hits in and out of the mix to keep things changing. Perhaps adding sub-bass frequencies only in a chorus or bridge section. Try bringing in the bassline in the middle of verse two, or adding a low octave double of the bassline somewhere along the way.

■ Experiment with acoustic spaces you have available in your home studio. Bathrooms, hallways, basements, and broom closets can bring a unique natural ambience to recordings. Trading digital reverb for the air of a real space can draw attention to your recordings. Tom Waits, Frank Black, The White Stripes, and many others play with this idea. When you find a particularly cool sound, make it into a trademark of yours—brand your own sound!

Sound on Sound: Les Paul

If Sam Phillips laid the groundwork for rock n' roll, the multi-talented Les Paul became its architect. Born in 1915 in Waukesha, Wisconsin as Lester William Polfus, Les Paul's career stretched over the better part of the 20th century, seeing him almost continually active as a country and jazz guitarist, arranger, inventor, and sound engineer. His jazz recordings are a must-hear for any aspiring guitarist, and his contributions to electric guitar design were immortalized in the Gibson "Les Paul" model long before his death in August 2009.

In the context of this book, Les Paul's most important contribution is his development of multi-track recording. His inspired drive to multi-layer guitars, the technological experiments in his New Jersey garage studio, and his confidence to present these (at the time) futuristic recordings to record executives pushed the state of the recording art further ahead than anyone since Thomas Edison. If we all somehow stand on the shoulders of giants, then anyone recording music, be it with a four-track cassette recorder or a 48-track Pro Tools DAW, stands on the shoulders of Les Paul.

Instead of going into detail about Les Paul's modifications to an Ampex tape machine that made multi-tracking possible, consider that the technology was a *means to an end* for Les Paul. He was in search of sounds. In order to create the sounds he imagined, he experimented endlessly. Modern studio techniques attributed to Les Paul include:

■ *Multi-track recording.* Or as he initially called it, "sound on sound" recording. With this came the sound of multi-tracked guitars, which have graced the recordings of every significant rock artist. This naturally led to multi-tracking vocals and other instruments, an essential studio-recording trick.

■ *Tape echo.* As noted earlier, Sam Phillips's rock n' roll recordings in the early 50s prominently featured a slap-echo effect generated by a tape machine, but Les Paul's innovations were light years ahead, and actually discovered several years earlier.

Les Paul noticed that the echo effect could be manipulated in several ways. By varying the tape speed, the placement of the playback head, and the amount of the playback signal fed back to the record head, Les was able to control the delay time and the regeneration of the echoes (repeats). On a modern digital delay plug-in, those are still the basic parameters available—delay time and number of repeats. You can hear Les Paul's amazing guitar tone and playing on his famous recording of "How High the Moon." Tape-based echo units like the Echoplex tape echo (or a modern remake like Fulltone's Tube Tape Echo) are still prized for their particular warm tone, and the subtle, unpredictable variations that a mechanical unit adds to the sound.

- *Varying tape speed.* First featured prominently on his album "The New Sound," Les Paul also pioneered the manipulation of tape playback speed to create new tones. By performing to a recording running at half-speed and then playing it back at normal speed, Les created sparkling, space-age guitar tones, which initially baffled other players. Once the secret was out, other artist began experimenting with recording and playback speed to alter the tone of guitars, drums, vocals, and even entire mixes. The Beatles' use of this effect on vocals is one of the better known examples, and the beautifully ominous edit in "Strawberry Fields" may be the most famous example of this manipulation; by varying the tape speed, George Martin and co. were able to paste together two takes of "Strawberry Fields" that were initially in two different keys and tempos! The shift in tone at that point in the final mix is one of the great "goose bumps" moments in recording history.

- *Close mic'ing.* By engineering many of his own recordings, Les Paul also experimented with non-traditional mic placement. Before there was multi-tracking, each microphone generally had to pick up more than one performer, forcing the engineer to place the microphone at least a small distance away from the source. In one sense, this can be good, creating a setting or space in the sound of the recording. However, placing a microphone close to the source makes the final recording sound like it is coming from inside the radio and is popping right out at the listener. Aside from their excellent performances of popular jazz songs, this method added a very present sound to the Les Paul/Mary Ford recordings, which certainly aided their success. Once this "in-your-face" sound hit the airwaves, everyone wanted to sound that way, and yet another modern studio technique was born.

- *Equalization as an effect.* To compensate for the loss of fidelity as the signal passes through electronic circuits and onto magnetic tape, equalization was normally used to correct this loss, and balance the tone. Les Paul began more freely applying equalization to shape sounds, especially when layering multiple guitars. George Martin and The Beatles would later take this to the next level.

- *The concept of "home recording."* His endless experiments with tape decks and eventual long performing tours led Les Paul to create portable recording devices that he could use at home and take on the road. Before this, recording was always done in professional studios with serious-faced engineers at the controls. By the 1960s and 70s, the Rolling Stones and Led Zeppelin had adopted this concept, taking mobile recording units to unique locations to capture new sounds for their records. Consider Led Zeppelin's famous "When the Levee Breaks" drum sound, which was recorded in a stairwell! The Red Hot Chili Pepper's breakthrough album "Blood Sugar Sex Magic" (produced by Rick Rubin) was recorded in a haunted California mansion—how's that for home recording? Without Les Paul's concepts and innovations, musicians might still be shackled to the big studios today.

As a true multi-talented genius, Les Paul should remain an inspiration to musicians and non-musicians alike. He changed the technology and the music as he saw fit, turning his ideas into realities; the ultimate role model for the creative person. He also strived to remain curious and young at heart, continuing to perform music all his life.

Along with Sam Phillips's "imperfectly perfect" concept, you should also think like Les, putting raw ideas first, and trying to bend and shape sounds to fit your imagination, forging new tools for the job if necessary.

Phil Spector

One of the most fascinating personalities in music recording, Phil Spector's production genius and penchant for threatening musicians with firearms has earned him a place in music history—and a monogrammed bunk in the same prison as Charles Manson. His intense personality led him to push performers to the brink of exhaustion, recording take after take until he felt the music was perfect. Those who didn't want to play along often found themselves staring down the barrel of a pistol; an unusual habit of Spector's that eventually brought an end to his career via incarceration.

Despite his well-publicized murder conviction, Spector's legacy of music production and studio innovation will likely remain his dominant image in our cultural memory. From the early 60s through the 70s, his "Wall of Sound" recording concept and studio per-fectionism produced hit after hit, including the most-played song of the 20th century, "You've Lost that Lovin' Feelin'." The "Wall of Sound" brought success to many period artists including The Ronettes, The Crystals, The Righteous Brothers, Ike and Tina Turner, The Beatles, John Lennon, and George Harrison.

Spector understood the need to stand out on AM radio (are you seeing a pattern here, students of recording?), developing a heavy-handed recording concept to push his artists through the radios of the world. This "Wall of Sound" begins with a large group of musicians in a relatively small room. Four or five guitars, up to three drummers, doubled and harmonized basslines, strings, horn players, multiple pianos, and myriad percussion effects were all recorded in the same room, literally saturating the air with sound. This method of layering sound takes advantage of the complex interaction and blending that takes place when multiple instruments perform the same part live. As opposed to over-dubbing instruments one after another, this technique provides the particular magic that Spector was after. Additionally, instruments were fed into the studio's echo chamber (many of his recordings were made at Gold Star Studios in Los Angeles), adding even more depth and density to the sound.

Spector surprisingly rejected stereo sound, preferring mono recordings, which he claimed gave him more control as a producer. He also preferred not to roll tape until he heard exactly what he wanted to hear coming through the studio monitors. His philo-sophy was that it "doesn't matter what's going on in there (the studio), the only thing that matters is what's coming out of the monitors." His unique approach to using the studio as an instrument and pushing performers to the limit of their abilities made re-cords that sounded like no others up to that point in time, raising the bar once again for all artists and producers.

Food for thought: How can home recordists create a "Wall of Sound"? Consider these points:

- *Keep in mind that the parts were simple.* Most of the hooks and melody lines in Spector's big hits were very simple—it was the huge sound that made them stand out. Consider keeping your ideas simple, or perhaps creating one hook from a simplified fragment of melody or bassline that you can huge-ify using the following ideas.

- *Kidnap a few extra musicians.* Polish up your six-shooter, and see if you can draft a couple more guitarists to simultaneously play the rhythm guitar parts live in the same room. You can use spot mics on each one, but be sure to add in an ambient mic to pick up all at once—this might be the ticket to that fat sound. On second thought, skip the pistol, and try it with a case of beer. That'll work just as well with most guitarists. Try red wine on pianists, and whiskey on bass players. For drummers, reconsider the six shooter.

- *Simulate performance.* If you don't have access to (or space for) multiple players at once, try to build up your own wall. Simply overdubbing several instruments with single spot mics may not do it, so try this. Let's say you're overdubbing guitars to layer up four of the same part. Track the first one, then play it back *over a speaker, and at about the same volume* while you track a second part. Just use your studio monitors for this! When the second is tracked, play those two back while adding a third in the same way. What this will do is bring about room saturation, which is part of the "Wall of Sound."

- *Simulate space.* The next step after layering those parts is to add a chamber-type reverb to the sound, bringing it up to the next level. The best way to do this is with a convolution reverb plug-in, such as TL Space or Altiverb. Alternatively, use a digital delay that simulates a chamber, blending it in and adjusting the pre-delay until it fits the mix. Trust the monitors! It's all about what you hear. I'll get into reverb settings in Chapter 8.

- *Don't forget the percussion.* The key to the fat sound is the little details, like tambourine, bells, splash cymbals, maracas, and so on. You should have a box of toys around for this purpose, and if you don't, raid the next flea market you can, tapping and shaking everything in sight—kid's toys, pots and pans, wine glasses, and more. A handful of rice and/or beans in an empty Gatorade-mix container makes a killer shaker. Or spend a few bucks at the music store. Just play back your track and try adding a few accents here and there. Layer these via overdubbing and the speaker trick, and add some chamber 'verb.

The "Wall of Sound" is not everyone's cup of tea, but if a track is sounding limp despite a good hook and melody, consider giving these ideas a try. You may not even want to do this on the whole track; it could be a great way to make a chorus or even just a bridge stand out.

Motown

During the same time period as Spector's "Wall of Sound" success, the producer-kings of Soul music were pumping out a stream of hits from the aptly named "Hitsville U.S.A." in Detroit, Michigan. The Miracles, The Supremes, The Four Tops, Martha and the Vandellas, The Temptations, Stevie Wonder, Marvin Gaye, and numerous others had over 100 hit records on Berry Gordy's Motown records label in the 1960s alone!

The key to the Motown sound was the songwriting and production team of Holland-Dozier-Holland, who have since been inducted into the Rock and Roll Hall of Fame. Lamont Dozier and Brian Holland generally composed music and arranged the songs, while Brian's brother Edward Holland, Jr. composed lyrics and arranged vocals. This team was responsible for a majority of Motown's hit songs, including the radio staples "Heat Wave" and "Nowhere to Run" (Martha and the Vandellas), "Baby Love" and "Stop! In the Name of Love" (The Supremes) and "Reach Out I'll Be There" and "Standing in the Shadows of Love" (The Four Tops).

The musicians who performed many of these smash hits call themselves "The Funk Brothers," and included the legendary James Jamerson on bass guitar. Besides being masters of their instruments and the musical style they played, they knew how to *support the song,* creating a full sound while leaving the right space for the vocal performances.

Tip: The film *Standing in the Shadows of Motown* finally brings the proper recognition to the musical talent behind the Hitsville U.S.A. hit-making machine. Their anecdotes and inside look at the workings of the studio is a must-see for anyone in the recording business, and serves as a warning to those with less business savvy to watch out for your financial interest in the work you do!

With such a powerhouse of management, talent, songwriting, arranging, and production in the house at Motown, hit records were almost a sure shot. Nonetheless, there are a few typical sounds that came about from the engineering side that are important recording earmarks.

Motown was using three-track tape machines, requiring the studio's Chief Mixing Engineer, Lawrence T. Horn, to create sub-mixes of the rhythm tracks before further overdubbing could be done. Staff engineers would track the instruments without using signal processing, leaving this to the Chief Engineer. This policy led to a consistent sound in the mixes coming from the studio, making the Motown sound instantly recognizable.

The Motown sound had to be danceable. This meant that the rhythm track had to be prominent, which inevitably leads to mix problems—the vocals need to be clearly heard above a rhythm track that begs to be loud! Lawrence Horn used a clever trick to add clarity and thickness to the lead vocal recording while maintaining its natural range. By splitting the vocal signal to two channels, he could balance a natural sound with a processed one, using his experienced ear to fit the tone to the mix. One channel stayed in its unprocessed state, with all the natural dynamics of performance; the life and rhythm. The second channel he compressed heavily, and boosted the high frequencies (with a boost in high shelving or in the vocal clarity range around 5–8kHz). By blending in some of this processed sound with the original vocal, he created a bigger-than-life sound that pushed the lyrics out through those killer rhythm tracks.

Microphone technique in the Motown studios was relatively simple, using close mics or direct lines on the instruments while they all performed in the same room. Keeping the number of mics to a minimum helps to keep the sound clear of phase problems and unwanted reflections. Keeping in mind that tracks were being sub-mixed as they were recorded, the monitors were boss. The staff engineers were responsible for capturing the band's performance, and there was no chance to remix after the performance.

The luxury of so many tracks in the modern studio may actually be making musicians lazy, encouraging them to put off important decisions until later because they can fix it in the mix. Printing a great sounding mix to tape has the advantage of ruling out second guessing. Have you ever gone back and listened to a rough mix you made, wondering how the magic got lost somewhere along the way? Well, try the Motown mix sometime—create a two- or three-track sub-mix of all the rhythm instruments. Strike while the iron is hot! Import that sub-mix into a new session, and overdub to that. If you did it right, you'll be starting off with bit of magic to carry you into overdubbing. In the worst-case scenario, you can import the individual rhythm tracks and remix. Be sure that you have a good monitoring situation set up for this technique.

All in all, the most important piece of outboard gear at Motown was the trashcan. Inferior takes were thrown away and redone until the magic happened. Microphone technique at Motown was simple—put a good mic in front of a good performer—every recording started with an undeniably solid and inspiring rhythm track to which singers could pour out their hearts. What else is there to say? Strive for excellence in your performances, and the recording will show it.

The Beach Boys

In May of 1966, The Beach Boys released an album that broke from their typical "sun and fun" light, white rock n' roll formula. It was a product of producer/songwriter Brian Wilson's conscious break from the conventions of touring and recording. Over time, the album "Pet Sounds" has come to be known as one of the most influential rock albums of all time.

As rock n' roll corroded the idealism of 1950s American life, new affluence, technology, curiosity, and world views became the elements of a reaction that transformed music and society by the early 60s. Armed with electric guitars, drugs, record players, bell-bottoms, and a general distaste for things old and conservative, artists and producers began searching for new sounds on both the musical and technical sides of recording.

Brian Wilson, the bandleader, arranger, producer, and chief songwriter of The Beach Boys had separated himself from the regular touring and recording schedule of The Beach Boys in 1964 following a nervous breakdown. He began experimenting with meditation, LSD, and digging around in sandboxes for hours on end—any of which is likely to change one's perspective on the world, but which in combination can lead to annihilation of the ego and radical changes in musical perspective... But before you go running off to Home Depot for some 2×6s and five hundred pounds of sand, let's see what you may be able to learn from this groundbreaking recording.

Considering his previous success with The Beach Boys, a departure from their typical production formula was quite a risk, one that required vision and self-confidence to follow through. Wilson hired first-call session players, and recorded in top-notch studios in Los Angeles. After recording the basic tracks, Wilson bounced them to one track of (at that time) the new eight-track tape machines, offering him the additional tracks he needed to create the thick, layered sound he envisioned.

Brian Wilson's production aesthetic for "Pet Sounds" was strongly influenced by Phil Spector and The Beatles (their "Rubber Soul" album in particular). Wilson was fascinated by the wall of sound, and the amount of creative control it provided the producer for delivering one's musical vision to listeners. Going beyond Spector's focus on single songs as hits, Wilson took a cue from The Beatles "Rubber Soul," composing songs with interrelated meaning and a unified sound concept. Despite Wilson's amazing talents for songwriting and arranging, he also had influences, and stood on the shoulders of giants for a glimpse over the horizon.

If you're not familiar with the album, it is worth investing in a copy. Have a listen, and see if you can determine which instruments are being used in any given song. Then 23-year-old Wilson's genius shines in the arrangements alone, but his visionary feel for combining multiple instruments, voices, and effects to create new sounds is his unique contribution to the history of rock music production. The Beatles have confessed that without "Pet Sounds," there would never have been "Sgt. Pepper's." In this respect, Wilson expanded on Phil Spector's "Wall of Sound" concept of using the studio as an instrument. He seemed to be reaching for the type of sounds that would first become easily available with the development of synthesizers.

To top off the exemplary songwriting, production, and recording innovation, the vocal recordings on "Pet Sounds" have a performance quality and timbre that was like nothing ever recorded before. Wilson slaved over his own vocal takes, and pushed the others hard while recording background harmonies; a take was not kept unless it was flawless. Although this type of perfectionism is not called for in all recording scenarios, Wilson knew that the vocals on "Pet Sounds" needed to transcend all in order to reach the level of expressiveness and sound quality he envisioned. Particularly stunning is the remixed version of "Pet Sounds" with all instrumentation removed—it is a must-hear for every recording enthusiast!

Food for thought:

- *How polished do you want your sound to be?* When beginning a re-cording project, consider the sliding scale between trashy and polished. For a modern punk/rockabilly project, polished is probably not what you're after. Be serious about this! In this case, aim to capture first takes and off-the-cuff performances. Relish the ratty edges, and don't waste time sterilizing things with editing and auto-tuning. Aim to capture energy and excitement, and leave all the smooth production concepts alone. On the other hand, investing time and energy in the pursuit of that magic performance supported by precisely performed and recorded backing can take the music to another level. This can, however, be a painful process. Pushing yourself—or even more difficult, others—to perform that "perfect" vocal take may require hours of repetitive work, and can put a strain on the best of friendships! Careful logging of earlier performances will help you compare takes and feel assured that you are approaching your goal. As the producer/ engineer, try to estimate the time needed to engineer and perform the tracks you are aiming for, and leave ample time for experimentation; there's nothing like time pressure to ruin the vibe.

■ *Go after the sounds in your head.* "I need to hear the sound of electric butterflies feeding on flowers of chocolate and linen." "You got it pal, lemme get out the Tesla coil and leftovers from my eighth-grade chemistry set'... Okay, take five." Before reaching for a synth unit or some plug-in simulation device that 1,500 other engineers are using at the same moment, consider combining real sounds to create what you need. This doesn't mean that you can't use plug-ins at all, just consider other ways to use them. For the electric butterflies, how about pitch-shifting the chime of a half-filled crystal wine glass into the notes of a scale (or using water level to adjust pitch), and doubling this with fuzz-box guitar? Or piano. Some saxophone with the attack edited off may be the chocolate you need, or perhaps vocal humming. You can play the whole thing back through an amp, smash it in a compressor, or feed it into an echo unit to bring it all together... It's up to your imagination, and you never know what you'll find.

■ *Aim for a great vocal sound, no matter what the case.* A great vocal sound is one that fits the track best. In the case of the punk-a-billy band, this may be a Shure SM-57 held in the singer's hand, and later fed into a guitar amp for some more dirt. In any case, try doubling the lead vocal, the de facto standard for a solid sound since the 60s. You may only want to bring the double in on choruses or just on particular lines, but it is usually worth a try. For that thick "Pet Sounds"/Beatles vocal tone, the only way to go is a line-for-line, note-for-note perfect double- (or even triple-) tracked lead vocal. It's not easy, and was one of Lennon and McCartney's most hated studio tasks, but ultimately creates a sense of depth and dimension that is otherwise impossible to achieve. Unfortunately, just copying the lead vocal track and blending it in again creates such a perfect double, as to not create this effect—it just makes the vocal louder.

   Check out the CD-ROM for some processing tricks used to "thicken" vocals.

The Beatles

With volumes already written about The Beatles, their songs, recordings, personal lives, and on and on, it is pointless to attempt to write something new about the topic.

Instead, let's take a different view of the development of The Beatles studio recordings, and stand in the shoes of their engineers. When working as an engineer, it is your task to find the technical solutions to the often rather esoteric requests of producers and musicians. If you prefer to be the idea person, please don't seat yourself in the engineer's chair. There is no one more difficult to work with than a frustrated musician acting as an engineer, and no better audio engineer than someone focused on the task and prepared to be part of a team effort. Setting ego aside is key, and can be difficult with all these "artists" hanging about, indulging their delusions of grandeur. Grit your teeth, stay focused, and deliver these clowns something they can trumpet around town. But I digress...

In the early Beatles recordings (let's say, anything pre-"Rubber Soul"), the audio engineering was strictly the realm of the lab coat–wearing trained professionals behind the glass doors of the machine rooms. The roles were clear; the musicians performed, the producer directed the show and was responsible for delivering the recordings on time and within budget, and the engineers pulled the levers and turned the cranks. There were rules for everything in those days, and the engineering department was not open to suggestions from the guests.

As The Beatles garnered some success, they began to test limits. They wanted more bass on the records, as they were hearing on new recordings coming out of the U.S. They wanted the drums to have more impact, wanted to layer guitar parts and background vocals as their contemporaries were doing. All these suggestions were initially met with resistance in the studio. Who were these boys telling the experts how to do their jobs!?

Well, it happened that these were well-considered requests from a band who had earned their chops playing clubs, recording, and promoting themselves for quite a few years. They didn't only imagine sounds they might want to hear, but heard things on import records—sounds they were excited about, but couldn't explain. This led them to begin experimenting in the studio, which turns out may have been as great a talent as their pop songwriting skills.

Producer George Martin factored prominently into this creative equation. Martin had to deliver professional sounding records, regardless of what experiments might be tried in the studio. Ultimately, he would have to shoulder the blame if experiments were to fail (they didn't!). He had the influence to get the engineers to begrudgingly try unorthodox recording methods. These engineers were union workers who had responsibilities to the recording company, and were not motivated to bend any rules. A lucky turn of fate at the beginning of the sessions for the "Rubber Soul" album saw trainee-engineer Geoff Emerick step in as engineer. Emerick's lack of experience left him open to the influence of the artists, and he rose to the task, experimenting with microphone methods while all the time under pressure to record sounds of the highest caliber.

Martin was also well versed in classical music, and knew how to arrange instruments The Beatles knew little about, allowing them (and encouraging them) to integrate new sounds into their arrangements. He was a sort of father to the band, someone to be rebelled against, but Martin himself pulled off a few recording experiments—sometimes in the absence of The Beatles—which have since gone down in recording history.

With this overview of the situation in mind, consider these studio anecdotes and how they changed studio technique. Think about the way they approached turning ideas into recorded tracks. The next time you feel bored with the same old sounds, think about how you might approach a studio task differently:

- The recording of "Tomorrow Never Knows" found engineer Emerick in the hot seat. John Lennon wanted his voice to sound, "as though I'm the Dalai Lama, singing from the highest mountaintop, although I want to hear the words I am singing." (Lewisohn, 1988) Emerick disassembled a Leslie speaker, rewiring it to play the vocal through the circuit, a probable first for this particular effect. Considering the loads of gadgets found in the modern home studio, what's stopping you from finding odd new sounds? How about stringing two coffee cans together with a guitar string, playing a vocal into one can, while mic'ing the other? Then plucking the string in time with the song while recording? And blending this back into the mix? How about with reverse echoes? Consider the physical as well as the electronic—paper towel tubes, washbasins, rotating fans. Anything can become an effects processor; don't let your audio become trapped in the digital world.

- When recording the bass guitar for "Paperback Writer," Paul McCartney wanted the kind of deep, full bass guitar tone he heard on contemporary records. However, the engineers were reluctant to boost the bass frequencies beyond a couple decibels (dB). They always had to consider the technical aspects of creating a vinyl record from the tapes, and heavy bass could cause the stylus to jump. As a workaround, they placed a speaker in front of the bass amp, wiring it to act as a microphone. The output was naturally bass-heavy, and created the tone McCartney envisioned.

- One of the most popular recording tricks in The Beatles repertoire was the manipulation of tape speed to affect the tone of their vocals. By recording with the tape speed turned down, playback at normal speed shifted the overall spectrum of the tone up higher (overdone, this leads to the "Chipmunk effect"). The opposite is also true; recording the vocal (or instrument) with the tape sped up results in a deeper, darker tone when played back at normal speed. Modern samplers and recording programs allow musicians to do this. Many DAWs and production programs provide a "master tune" switch, usually set to A440 as a default. Try manipulating this and see what happens! A flabby bass sound could be recorded a few semitones lower with the tuning set down, and then played back at normal speed to "tighten it up." Pro Tools has a half-speed record function that allows you to create virtuosic guitar runs at the flick of a switch. Propellerhead's Reason allows you to easily manipulate audio, playing it back at different pitches in real time using their samplers. Even GarageBand allows you to retune loops and audio without changing the speed!

- ADT (Automatic Double Tracking) was developed for The Beatles by their engineers to get them to stop crying about having to double track their vocals. George Martin required that particular "thick" vocal sound that resulted from a singer performing the same vocal part twice, and it was a difficult task to sing the song note-for-note perfectly compared to the original vocal. The effect that real double tracking has is intense, and almost always makes a vocal sound instantly pro—it's just a lot of work. The engineers developed the technique of copying the vocal track to a second, synchronized tape machine, and then varying the tape speed slightly to create an "artificial" double; hence the term ADT. Using effects to "thicken" sounds is now commonplace, but has the predictable sound of effects. Try copying the vocal to another track, and applying a pitch correction plug-in to each track, however, use different settings on each track to create the illusion of doubling. Consider setting a MIDI controller to control an effect parameter in real-time, and get your hands in the mix. Also consider just doing it the old-fashioned way and doubling the part by performing it again.

- Pass the ketchup, please. Despite having one of the greatest rock n' roll voices *ever,* John Lennon was always dissatisfied with the tone of his own voice, and in his frustration, once told Martin to "Change it, do anything! Put some ketchup on it!." As a result, many of his vocal tracks are loaded with echo, filter effects, ADT, tape speed effects, and the like. Many of you at home probably feel the same way—nothing sounds worse than your own voice playing back at you. Slap-echoes, filters, phasing, reverb, and the like, are all available at the touch of a button now, so what else can you do to relieve the pain? Try other ways of making your voice "bigger than life," starting by copying the track a few times. Pan copies left and right, and apply different EQ settings and filters. Make a stereo mix of the track, pitch shift the sub-mixed track up, singing to that, and then lower the pitch of the vocal back down into the regular mix. Blend this with the main vocal. Sing a double through a megaphone. Or call on the telephone. Stay up late partying, and then record a vocal first thing in the morning. Do anything to make that track stand out in the crowd.

- Challenge yourself to create something using a new instrument. Although when performing they were a four-piece, guitar-driven band, The Beatles' studio recordings include everything from harmonica to harmonium, sitar to classical strings, trumpets to tamboura, and every kind of percussion they could get their hands on (including knee slaps and foot taps). John Lennon once said that he could "make music on any instrument, give [him] a tuba, and [he'll] play you a song on it." This statement could be taken as a bit of grand posturing, but consider the inspiration that often comes out of picking up a new instrument. Those first notes you play have a certain sense of exploration that you don't get when playing around on your usual axe. Beg, borrow, or steal an accordion, sitar, lap steel, steel drums, or some other unusual (for you) instrument. Record your first take, and see what happens. Play from the gut, and don't let "mistakes" get in your way. You have a great excuse to be lousy at it—it's not your instrument! This often leads to great arrangement ideas and new melodies.

- Studio experiments in the 60s led to close mic'ing and manipulating sounds, something we have all heard done over and over since then. With all the plug-ins and processors readily available, it's very easy to search for sounds by using this library of tones. If nothing is tickling your fancy, try removing effects, and moving the mic away from the source. As iPods and other MP3 players with their in-ear headphones (ear buds) shove the music closer and closer to our ears, consider putting some space, distance, and air into your recordings. Take advantage of the spaces you have, and let things breathe.

Frank Zappa

Since the pop/rock innovations of The Beatles, there have arguably been few mainstream artists who have pushed the boundaries of songwriting, production, and recording in the genre. Frank Zappa, however, was the flip side of the coin. An amazingly prolific writer and performer, Zappa recorded over 80 albums, while consistently giving the middle finger to the establishment. These vary from polished studio gems, to live recordings and unusual cross-breeds wherein he would overdub studio tracks onto live recordings or even re-synchronize guitar solos extracted from live improvisations onto studio recordings to create complex rhythmic interplay, a technique he dubbed *xenochrony*.

Aside from his sarcastic wit and guitar virtuosity, Zappa was also a talented audio engineer driven to experiment with sound. I'll delve into some of his recording techniques as I get to individual instrument recording techniques, especially guitar and drums. Particular to Zappa's recordings is his idea of "conceptual continuity," meaning that concepts within his music expand beyond the song or even album. Conceptual continuity led Zappa to freely edit together elements of recordings he had made at different times to create new arrangements, again showing the power of a concept behind music production.

Brian Eno

Brian Eno could be considered a philosopher and musical explorer as much as a producer and musician. Some of his best-known productions include David Bowie, Devo, The Talking Heads, U2, British Indie-rockers James, and recently, Coldplay. Eno's production style favors the spontaneous and accidental over the composed and rehearsed elements of music, using unorthodox methods to arrange music and elicit performances. For example, during the production of U2's "Achtung Baby," Eno's job was to come into the studio every few days, and "erase anything that sounded like U2." He did this to challenge the band to explore new sonic territory.

Along with British artist Peter Schmidt, Eno developed "Oblique Strategies," a set of cards designed to aid self-criticism and problem solving in creative endeavors. When at a creative crossroads or in a mental block, the artist can draw a card, revealing abstract statements for consideration, such as "a line has two sides," "use clichés," and "go to an extreme, come part way back." There are now small applications for your computer that allow the drawing of a virtual "card," so that you can consider these strategies while working on your music. I highly recommend searching for "Oblique Strategies" on the web and adding one of these applications to your music-producing toolkit.

Eno has also worked on developing programs for "generative music," in which the resulting compositions may never be played back the same way twice. As opposed to the confines of recorded music, this concept brings the unpredictable element of live performance into the formula. He believes that "it is possible that our grandchildren will look at us in wonder and say: 'You mean you used to listen to exactly the same thing over and over again?'"

Food for thought:

FOOD FOR THOUGHT

- *Innovation is often the result of questioning concepts you take for granted.* We have considered the contributions of some innovators in pop and rock music, and a common thread has been the willingness to experiment and take risks to pursue a concept. Consider what you want to express (or if you are strictly engineering, what the artist wants to express) and attempt to turn this into a recorded sound. This will not always work. In fact, success will likely be the exception rather than the rule at first. Don't let this stop you, but go after these concepts whenever you can.

- *Consciously imitate that which you admire.* All of the example producers and artists did precisely that. Don't copy, but imitate; your own style will develop, regardless.

- *Invite criticism.* As we all spend more and more time communicating via the Internet, filtering out that which we find unpleasant, we avoid direct contact with others and criticism of our work. Present your rough mixes, and don't defend them; if you know the kick drum is too loud, let someone else confirm this—if that's their only criticism, you're doing well. Play the music for a group of people; it should make them dance, cry, laugh, or shake their bodies. Communicating a feeling with sound is successful music.

- *Take a break.* Be sure to rest your ears and mind after intense working sessions. Burnout happens regardless of the job, and your ears are your most precious tools. You've got a lot of ground to cover, and an open mind and fresh ears are all you'll need to begin learning how to record some great music.

2 The Basics of Audio Recording

In order to get a handle on the complexities of the modern studio, with multitrack DAWs, plug-ins, outboard processors, and a barrage of buzzwords coming at you from every direction, you need to first get down to the basics and clear some of the smoke surrounding audio recording.

Since the 1990s, when digital tape decks made it possible to record clean and clear audio in a home studio environment, gear manufacturers have not let up trying to sell musicians a new line of toys every season. This often leads them to believe that their recordings sound inferior because their gear is inferior. The gist is that you need to throw more money at the problem, when in fact, the gear made 10 years ago is more than adequate (in most cases) to make recordings you couldn't tell apart from recordings out of a "big studio." So let's learn how recording works, from the bottom up, and as you go, you'll learn the techniques and working methods that all add up to a serious improvement in the sound of your recordings.

Why Knowledge Is Power

The sound of your recordings will benefit most from the confidence that comes from being sure of what you're doing at every step.

A favorite expression in recording circles is "garbage in, garbage out." In other words, the sound source that you initially record is not going to magically change within the microphones, blinking boxes, and other technology of recording to somehow become better. Lazy musicians and producers may want to "fix it in the mix" if some part of the recording is not up to par. In this case, "garbage in, garbage out" is the proper response; a sloppy guitar part, unconvincing vocal take, or out-of-time drum fill can often be adjusted with plug-ins and editing, but there is no replacement for a great performance from the beginning.

Along with learning about the recording environment itself, a top-notch engineer or producer will learn about the instruments being recorded, and how they should sound. It's okay if you can't tune a piano! Just know enough about it to hear that something's wrong, and get someone who can tune it. Knowing the conventions of different instruments will also help you to experiment with alternate tunings, different performance methods, and unique effects to make your recordings stand out.

Understanding Sound Recording Basics

First of all, you'll need to understand the fundamental process of recording audio. When you understand that, then multi-track recording is as easy as, "just that idea, times the number of tracks you have." Learning about plug-ins, processing, and effects becomes simpler, as you'll understand *where in the chain of things* these happen—you will have a clear overview.

So let's start all the way at the beginning.

Edison's Wax Cylinder Phonograph

Although we won't be focusing on techniques used as far back as Thomas Edison's time, his wax cylinder phonograph still provides a clear example of how sound recording works. As noted, you can then relate to this simple concept to help understand the new and complex steps added to the process in modern recording. These sorts of analogies make things easier to remember, as visual memory is especially strong.

Thomas Edison's wax cylinder phonograph, as illustrated in Figure 2.1, functioned (in a nutshell) as follows. A sound collecting horn—literally a large funnel for the sound

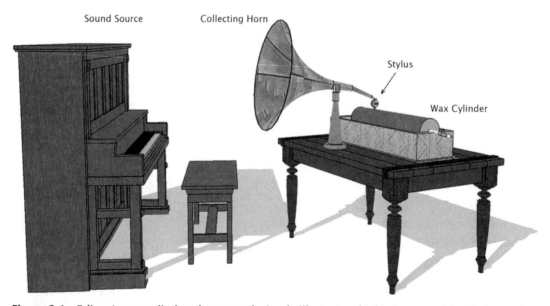

Figure 2.1   Edison's wax cylinder phonograph simply illustrates the basic concept behind sound recording. Sound vibrations in the air are translated into mechanical vibrations via a diaphragm. These vibrations are etched into a wax cylinder, and recorded for posterity.

waves—was aimed at the sound source. At the narrowest part of the horn was a thin diaphragm to which a needle was directly attached. Incoming sound caused the diaphragm to vibrate, moving the needle in a way analogous to the incoming sounds (note the word "analog" in there). The needle carved a representation of this sound into a rotating wax cylinder, thereby recording the incoming audio.

The sound could then be reproduced by placing the needle back on the rotating cylinder. The engraved sound representation moved the needle, vibrating the diaphragm into the horn, and reproducing the sound wave. Rather low-fi, but simple to understand!

As primitive as this was, this wax cylinder phonograph worked, and all audio recording since then is based on the same principle—a *transducer* (a device that transforms one form of energy into another; in this case sound energy into the energy of mechanical motion) is used to create a representation of the audio, saved on some physical medium. To hear the sound, the medium is called upon to trigger a series of motions and energy transformations that reproduce the sound.

Analog Tape Recording

If you jump ahead to the early days of analog tape recording, the concept remains the same, although technological advances far improved the fidelity of the recorded sound. In a nutshell, a microphone (containing either a metal ribbon or a flexible diaphragm) is placed near the sound source. The energy of the sound causes the ribbon or diaphragm to move, creating a small electrical current. This current is then amplified using a microphone preamplifier (sound familiar?). The changes in the electrical current cause a magnetic field to fluctuate. These changes in the magnetic field are recorded onto a moving magnetic tape.

In the playback process, the magnetic tape rolls past a playback head, which senses the recorded signal on the tape, and changes this back into an electrical current. This small current is then amplified with enough power to drive speakers, which are incidentally also just big magnets connected to diaphragms!

The process has become more complex in the search for better fidelity, but is still intricately linked to the basic concept of Edison's wax cylinder. Let's take one more leap, and you'll know how modern recording works.

Modern Digital Recording

We still use microphones as the first step in modern sound recording. Fact is, we must transform the physical sound waves into an electrical signal in order to record them, even when carrying out digital recording, there's just no way around this yet!

So what happens is almost the same as in the analog days; you just add a couple more steps in the interest of making things complicated . . .

The microphone turns the sound into a small electrical current. A microphone preamplifier boosts that signal to a useable level. This signal is then fed into an *analog-to-digital converter*. You can learn more about how this works later in this chapter, but for now it will suffice to say that the A/D converter turns the analog electrical signal into a digital representation that can be stored as a whole lot of ones and zeros (that's the form in which computers store information).

When you want to play back the audio, you need to turn all these ones and zeros back into an analog electrical signal, which can be amplified to drive the speakers, as usual. For this step, a *digital-to-analog converter* (D/A converter) comes into play; the stored information about the sound is converted into an electrical signal that approximates the original signal you put in. The current technological quest is to improve this approximation so it's as close to the original as possible!

Food for Thought: A/D and D/A Conversion   The quality of the A/D converters is an oft-overlooked, yet critical, element in the recording chain. The cost of top-notch A/D conversion is a price worth paying if you want not only an accurate representation of your sound source, but also a tone that is easy on the ears.

You may ask, "Is there so much to it, that the sound quality can differ?"

Yes, and this is the last step in getting your audio into the digital realm, so it's best not to skimp here. Inexpensive converters may suffer from clock jitter, harsh sounding anti-aliasing filters, inferior dither, or noise or disturbances from the power supply, among other things. Once you've turned your signal into ones and zeroes, your precious audio is safe, and it's at the best quality it's going to be.

For the home studio, investing in just a pair of great converters may be enough, because you will likely do a lot of overdubbing, for which you usually require two channels at a time, at most. Apogee, Lucid, and a few other manufacturers offer stereo units of top quality, and RME's Fireface model 400 or 800 provide four or eight channels of great-sounding conversion at affordable prices. If you think you'll need more than two channels at a time rather often, the RME units or Apogee's ensemble (for Mac) will let your carefully recorded tracks show what they are really made of.

When listening back to your audio, you will also want to be able to accurately hear what you have so carefully recorded. This is the *D/A conversion* (digital to analog)— the process of turning those ones and zeroes back into an audio signal.

As with the A/D conversion, D/A conversion quality can make a significant difference in your recordings by influencing how you hear your recorded audio. When it comes time to process and mix your tracks, your decisions will be based entirely on what you hear, and the more accurately you hear, the better you can EQ, compress, process with reverb and delay, and place your tracks in the stereo field. Furthermore, good quality D/A conversion will provide audio quality that is less fatiguing on your ears, allowing you to work more effectively.

Now, let's review a real-world situation—recording a person who is playing the guitar and singing. A typical modern setup might be to place one microphone in front of the guitar, and one at head level for the performer to sing into (more on microphone choice, placement, and technique in Chapter 5, "Microphone Technique"). Figure 2.2 shows a diagram of the situation.

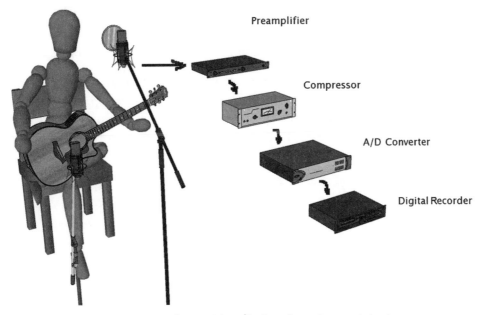

Figure 2.2   A modern setup of a musician playing the guitar and singing.

What's happening in Figure 2.2:

1. The performer's physical motions (strumming the guitar and singing) create sound. The sound travels through the space in which the performer is sitting.

2. The sound energy causes the microphone's capsules to vibrate, and the microphones (transducers!) transform this mechanical energy into an electrical signal.

3. The (now electrical) sound signal is processed as the engineer sees fit, and fed into an A/D converter. The A/D converter creates a digital representation of the electrical sound signal, which can then be stored on a hard drive or other digital storage device. The sound has been recorded!

4. Whenever the engineer wants, the recording can be played back. The first step is to convert this stored digital recording back into an electrical signal. This is the D/A converter's job.

5. The signal can be further manipulated either in the digital realm (plug-ins) or in the analog realm (outboard gear) on its way to being amplified so you can hear it.

6. The electrical signal is amplified to a level strong enough to power the final set of transducers—the studio's monitor speakers. The electrical signal has now been changed back into mechanical energy that you can hear. You've come full circle; from the mechanical energy of performance, to an electrical signal from the microphones, converted to digital and saved, converted back from digital to electrical, and finally back into mechanical sound waves that you can hear.

All the Little Pieces Add Up

This is a very straightforward (if a bit simplified) way to understand the recording process. The truth is that many aspects of the recording environment and setup can radically influence the sound and people's impression of it along the way. The main points I will focus on in coming chapters, which will help you improve your recordings, are the following:

■ Room setup and acoustics

■ Microphone technique

■ Equalization/compression

■ Other signal processing

■ Mixing/mastering tips

So, once you've learned some of the tips, tricks, and techniques in this book, here's what you want to do with the singer/guitarist pictured in Figure 2.2:

■ *Listen.* First of all, you'll use your ears—in the room with the performer—and consider the voice, instrument tone, tuning of the instrument, and the room acoustics.

■ *Choose microphones.* You'll pick out the right microphones for the task, based on what you hear.

■ *Place the microphones.* Mic placement is as important as mic choice. You'll learn where to find the likely *sweet spots,* that is, where the instruments sound best, so you can then make a test recording and adjust as needed.

■ *Create a signal chain.* You'll pick a fitting preamplifier and, if necessary, compressor, equalizer, or other outboard gear. You'll know how to connect them and optimize the signal path for the best possible sound (which may even include some tricky gear manipulations!).

■ *Set tracks and monitoring.* Once you get that beautiful audio fed into some tracks, you'll usually need to provide the performer with a way to hear it. Monitoring is one of the make-it-or-break-it steps in recording. The talent needs to hear the song to their liking, period. I have seen many sessions fall apart because of poor monitoring, and this is usually where the vibe gets bad and the finger pointing starts. This is not always just a technical issue, and a good producer will know how to handle this.

■ *Test and troubleshoot.* "Are we ready to roll?" You'll be able to get set up and get a good sound in a short time, and feel confident about what you're doing. Whether you're just engineering or wearing all the hats, solid technique will serve you well. Always.

■ *Play back those tracks.* Your control room monitors, acoustics, and ergonomics are key to being able to listen and make critical decisions. Not only do you need good monitor speakers, but you also need to be sure the acoustics of the room are not skewing your perspective. Beyond that, the room should be comfortable, conveniently laid out, and simply a groovy place to sit back and listen to what you've recorded. If you can't sit comfortably and hear clearly, the most important step might be missing—hearing what's been recorded!

It's thrilling to play around, experiment, and get some far-out stuff happening in the studio once everyone's on the right wavelength. But the truth is, there always has to be someone who knows how to keep the technical side running. If you practice solid technique, it will be there to save you when things get interesting...

As you get a handle on these steps and learn the tricks of the trade, you'll be recording tracks that you can be proud of when it's time to mix. The devil is in the details, and if he isn't then we'll find a way to squeeze him in.

Tip: Please review the singer/guitarist example illustrated in Figure 2.2 if you are still not clear on how the audio moves through the recording system. Use the figure to get a picture of it into your mind's eye, because this kind of visualization will help you again and again. In fact, basic troubleshooting in the studio relies on being able to think through these steps in the recording process. If you can visualize the audio process, you will be well on your way to becoming a capable engineer.

INSIDE THE BOXES
Some Studio Lingo for You   The basic recording process, whether it's just a singer with a guitar, a four-piece band, or an entire orchestra, is usually referred to as *tracking,* meaning recording the initial tracks you're going to work with. If the same track is recorded a few times, refer to each one as a *take,* such as Vocal Take 1, Guitar Take 3, and so on.

Adding more recorded audio to those tracks is called *overdubbing.* The word *dubbing* originates from movie production, and refers to the process of replacing speech in a film production. The term has been adopted by audio engineers, and now refers to the process of adding any new part to previously recorded audio, probably because the voice was the most likely instrument to be replaced after recording a live band. Nowadays, musicians love to pile on guitar parts, walls of synth sounds, loops, percussion, and mysterious AM radio transmissions received in the wee hours of the morning. At least I often do.

Once you start adding overdubs, the initial recordings are usually referred to as the *basic tracks* or the *basics*. Since you may have the same musicians recording several parts in your project, this helps clear things up when referring to a track, for example: "Hey mate, there's a strange sound in the basic guitar track; could you punch in that bar?"

Punch in is not what you and Fred Flintstone do every morning at the gravel quarry, but what you do when you need to fix a mistake on an otherwise good performance. The musician plays along while listening to the original performance. Shortly before the mistake, the engineer begins recording, and the musician continues playing, hopefully performing the part correctly this time. Shortly thereafter, the engineer exits recording mode (*punches out*) and then stops playback. The time before the *punch in* is called *pre-roll,* and allows the performer to find her place and get into the performance. After the engineer exits record mode, it is good practice to allow at least a few seconds of *post-roll.* Post-roll playback allows the performers to hear if they matched the performance well, and also spares them what can be a rather odd, unmusical cut off of the music after getting into the groove. This helps subsequent takes flow more easily.

Tip: Come up with some *shorthand terms* for labeling tracks, and stick with them. Some examples are "vox" for vocals, and "gtr1" and "gtr2" for different guitar parts. This will help you stay organized and find files quickly as you work. I find it useful to use capital letter abbreviations for the song title, then the part, then the take. This makes it easy to find an audio file any time. Example: For a song called "It's Been A Long Time," the second take of the vocal harmony could be labeled "IBALT_vox_harm_take2." This allows you to find backups easily, should something happen to your data. I've often made the mistake of altering a file irreversibly, and needed to find the backup to save my butt. At times like those, you'll be very happy that the filename is not "Audio2." After just a few months of recording, you might have the file "Audio2" showing up 12,673 times on your hard drive . . .

Once you've seriously jammed up the pipes with your artfully arranged audio tracks, it's time for *mixing,* which is the task of balancing all the parts relative to one another, featuring important elements, and creating a sense of space, all the while keeping in mind that things should sound clear and convey the message of the song. That wraps it up nicely in two sentences, but I'll get down to the details in Chapter 11, "Mixing: Balancing Art and Craft."

After mixing and before duplicating the music for distribution comes the elusive and mysterious process of *mastering.* Mastering is the final overall adjustment that a stereo (or in some cases surround) mixed track goes through before being released.

Highly paid professionals lurking in tightly sealed buildings can make your mix louder, clearer, and most importantly make it *translate well onto almost any playback system,* from iPods to boom boxes to those little crusty speakers in elevators. There are entire books on mastering, and if you plan on mastering someone's music, I suggest you sleep on a stack of them. I'll go over some basic tips and tricks to get you off on the right foot in Chapter 12, "Mastering: The Final Adjustments."

3 The Most Important Tool: The Microphone

The basic tool of recording is the microphone. Throughout the last century, it has been the one constant in the world of recording audio. The format in which the recorded audio is stored has advanced from a needle stylus carving into wax, to magnetic recording heads rearranging iron particles on flexible tape, to endless strings of ones and zeros being recorded on a computer's hard drive. In comparison, basic microphone design has barely changed since the 1940s! In fact, recent years have seen the bold return of one of the older forms of the microphone—the ribbon microphone—with the most sought-after models being either 60+ year old vintage mics or reproductions thereof.

As you learned earlier, a microphone is a *transducer;* its purpose is to transform the mechanical energy of sound waves into an electrical signal that you can then store as a recording. There are, however, a few different basic technologies used in microphones. Learning about microphone types and their characteristics will help you make educated microphone choices when recording, and will help you capture sound in a skillful and creative manner.

Microphones generally fall into one of three groups, based on the technology used for the transducer:

- *Ribbon microphones.* A thin strip of aluminum is suspended between the poles of a magnet. When a wire moves in a magnetic field, an electrical voltage is generated, so the ribbon's sympathetic vibrations act like wires and create electrical signals that are directly related to the source sound. Physics *can* be pretty cool!

- *Diaphragm microphones.* These are the most widespread type of microphone. They use a circular membrane that transforms the sound vibration in the air into mechanical vibrations in the membrane.

  The way in which the diaphragm's motion is turned into an electrical signal varies as well, and is described in more detail later. Sub-types include moving coils, condensers (also called capacitors), electret condensers, and even the occasional carbon microphone—which is the technology used in old telephone handsets.

■ *Piezoelectric crystal microphones.* With these types of microphones, sound vibrations compress a crystal, which creates an electrical signal.

Note that there are other interesting microphone designs, including ones using fiber optics and lasers, but due to their limited (if any) application in studio recording, I don't get into them here.

You will often see words like tube, FET, active, and so on, also associated with microphones, but these refer to aspects of the microphone's electronics, and not the basic transducer type. These aspects may certainly influence the sound of the microphone, but the basic type usually most influences its overall character. Let's start off by learning about each transducer type, and then the chapter will investigate some of the aspects of microphones in general, and the way they influence sound.

Ribbon Microphones

Try this—go and get a piece of plain white printer paper. Now hold it up in front of your face, fingers lightly touching the back of the paper. Now just hum, speak, or sing a tune. Do you feel the paper vibrate along with the sound? Now imagine that instead of a piece of paper, which is already rather thin, you were to speak into a strip of aluminum tape just a few microns thick. For comparison, a spider's web is about one micron in thickness. Imagine how well this thin sheet would pick up the vibrations of sound! Now, suspend this strip of aluminum between the poles of a strong magnet, and *presto!,* you have a ribbon microphone. Figure 3.1 shows a typical ribbon mic.

INSIDE THE BOXES
Understanding Frequency Response Microphones often have characteristic "bumps" or "dips" in their frequency response due to microphone design and diaphragm tension. This is the difference between the input sound and the microphones output at different frequencies. For example, ribbon mics are often more sensitive to low frequencies than to high frequencies. The result is that recordings sound "bottom heavy" and may need to be corrected with equalization. You could, however, choose a bottom-heavy microphone on a sound source that has an unusually thin or "trebly" sound, such as a small-bodied acoustic guitar, or a thin snare drum. The additional sensitivity to the sources low frequencies helps to bring out this range of the sound, resulting in a more balanced tone. By choosing a mic with a fitting *frequency response,* you can sometimes bypass the need for EQ later. Because less processing leads to more solid-sounding tracks, taking the time to choose the right mic will make your recordings sound better.

Figure 3.1   The AEA R84 is a top-notch replica of the venerable RCA 44 ribbon microphone. Its sound is characterized by a warm, extended bass response and a clear treble response. It excels on vocals, drums, and stringed instruments.

The technology behind ribbon mics is very simple, and creates an amazingly detailed reproduction of the sound source, hence its continuing popularity, even today. Ribbon mics have an inherently smooth *frequency response* at higher frequencies, which many engineers find to have a pleasing sound, especially when recording to a digital format.

On the downside, ribbon microphones have a very low output level, and need to be coupled with a microphone preamplifier that can cleanly amplify their signals to useable levels. Luckily, many modern mic preamps are up to the task.

Additionally, there are two Achilles' heels built into ribbon mics, and you must be careful not to destroy these delicate instruments:

■   *Moving air* can destroy the ribbon. This includes puffs of air from speaking or singing the letters B, F, P, or T (often referred to as *plosives*), air from fans or air conditioners, and possibly even air currents created when moving the microphone. Be careful by using pop-screens when recording vocals and by covering the mic when moving it or when it's not in use!

- *Phantom power* (48v power applied via the mic cable; necessary for most capacitor mics) can destroy the ribbon as well. Electrifying that poor, skinny little aluminum strip can ruin it for good, so be aware of the characteristics of any ribbon mic you are using. Some actually need phantom power for their on-board electronics, but most classic models must never be connected to phantom power.

Additionally, you should store ribbon mics upright, not lying on their sides, as the ribbon can sag under its own weight over time.

Despite the need for a bit of caution, ribbon mics are robust and long lasting, and provide a uniquely detailed sound that excels in recording the fast attack of percussive instruments and plucked strings and the detailed sound of the human voice. They are especially useful when recording bowed instruments (such as the violin and cello), because their smooth frequency response helps avoid accentuating harsh tones from the bow.

Well-known examples of ribbon microphones include the RCA Models 44 and 77, Coles 4038, Beyer M-160, Royer R-121, and the whole line of top-notch reproduction ribbon microphones from AEA (Audio Engineering Associates). This is by no means a complete list, but these models are rather common in the pro audio world, so you may want to note some model names so that you'll feel comfortable with some of the typical studio jargon! A quality ribbon microphone usually costs $800 or more, but there are also many inexpensive ribbon mics available from Nady, Golden Age, CAD, Oktava, Cascade, and a few other companies. Upmarket mics use higher quality parts, which in many cases improves the sound quality, but also costs a pretty penny (the transformer in pricier mics is typically the expensive, higher quality part). Since the technology behind ribbon microphones is relatively simple, these "budget" mics provide excellent value for your recording gear dollar, most falling between $100 and $400. The Cascade FAT HEAD and R-84 style mics from Nady and Golden Age are project studio favorites that hold up well when compared to their pricier competitors.

Diaphragm-Based Microphones

Far and away the most widely used microphone design involves a thin, circular membrane (now usually made of synthetic material) that vibrates sympathetically with the sound source. The circular diaphragm design offers more durability than the thin strip of aluminum used in ribbon mics, and tensioning of the diaphragm allows designers to tailor the frequency response of the microphone. For example, a rise in high-frequency sensitivity may be desirable in a microphone for recording dull-sounding sources. See Figure 3.2.

Figure 3.2 The Shure SM-57 is one of the most common diaphragm-based microphones. Its sturdy construction and reliable performance have made it a go-to mic in recording studios, broadcast, and live sound reinforcement all across the world.

The three most common designs for diaphragm-based microphones are:

- Dynamic microphones
- Condenser (capacitor) microphones
- Electret condenser microphones

Dynamic Microphones

Dynamic microphones are the most common, least expensive, and most rugged type of studio microphone in use today. The ubiquitous Shure SM-57 and SM-58 (or other manufacturer's copies of this design) can be found in any studio in the world, and are a perfect place to begin your own mic collection. These mics sound pretty good on just about any source, and are inexpensive and sturdy. Engineers joke that, if the need arises, you can disconnect the mic cable, hammer a nail into the wall with one, plug it back in, and keep tracking.

Although there is a wide range of style and complexity within the large range of dynamic mics, their basic design is also rather simple. The diaphragm is physically connected to a coil of wire, which then sympathetically vibrates with the sound source and moves within the field of a magnet. Again, this gives rise to an electrical signal that can be further amplified and fed into the recorder. In fact, dynamic mics work like a speaker in reverse. If you place a speaker in front of a sound source, it will pick up the sound, the signal then appearing as an output at the speaker's input poles—a trick sometimes used in studios to create a unique recorded sound.

The nature of dynamic microphone design provides for a stronger output signal and a sturdier microphone, but a generally more uneven frequency response. Because of this, different dynamic mics are often chosen for particular jobs in the studio. Learning these tips is important to good studio technique, and you'll learn about this in Chapter 5, "Microphone Technique."

Popular dynamic mics include the ElectroVoice RE-20; AKG D12 and D112 (famous for kick drum mic'ing); Sennheiser MD-421, MD-441, and e609; and the Shure SM-57,

SM-58, the classic SM-5, and the SM-7. Again, this list doesn't even approach completeness, but is meant to familiarize you with a few more common models.

There are also many convenient and popular drum mic'ing sets of dynamic microphones from Audix, Shure, Audio Technica, and others that often have microphones tailored to the job of mic'ing toms, snares, and kick drums. They are very useful for live sound reinforcement or live recordings, but the highly directional nature of these mics may sacrifice some tone for noise rejection, and their pre-EQed sound may not suit recording purposes—important points to consider for studio work.

Condenser (Capacitor) Microphones

Condenser (capacitor) microphones (see Figure 3.3) involve a more complex design, with the designer's goal generally being a more sensitive (higher output) microphone with a wider frequency response and a faster *transient response* than a dynamic mic— that is, the lighter materials used in the design allow the sympathetic vibrations to move the diaphragm more quickly and accurately, therefore providing a better "picture" of the sound.

For you Propellerheads, the basic idea is that the diaphragm (which is metallic) is built as one plate of a charged capacitor. Sound vibrations change the distance between the two plates, thereby changing the capacitance, and this is manipulated to create an electrical audio signal. The theory behind this goes beyond the scope of this book, but it is interesting to note that a different technology is used, in this case not involving induction of an electrical signal in a wire in a magnetic field.

The famous Neumann bottle microphones made in Berlin in the late 20s were some of the first condenser microphones to come into widespread use, and Neumann's later model M49 and U47 microphones are still some of the most coveted pieces of studio gear around. Hundreds of manufacturers now offer condenser microphones, and prices vary according to the quality of the parts—gold-sputtered diaphragms, hand wound transformers, top-quality electronics, and famous brand names (!) can drive prices into the thousands-of-dollars range, although there are quality condenser microphones to be had at affordable prices.

Particular project studio favorites include Audio Technica's AT-4033, 4050, and the inexpensive AT-2020. Shure, makers of the ubiquitous SM-57 and SM-58, also make the KSM line of condenser mics, known for being flexible mics for any application. The Shure SM-81 is a very popular small diaphragm condenser mic for recording acoustic instruments. Neumann has, in recent years, begun offering more affordable condenser mics, such as the TLM-103 and the new TLM-102, which are an excellent investment for home engineers looking for a touch of that "big studio" Neumann tone.

Figure 3.3   The RODE NT-1a is an affordable, professional-sounding condenser microphone. Although it costs just $200, the RODE's sound quality holds its own with mics costing many times more, making it an excellent choice for the budget home studio.

There are far too many manufacturers to provide a comprehensive list of condenser microphones, so I make further suggestions in Chapter 5 on a source-by-source basis.

Food for Thought: How Much Is That Microphone in the Window?   You gear junkies out there have probably scratched your noggin many times over the wide price range of large diaphragm condenser microphones. In this sidebar, I consider some of the most common factors in determining the price of the microphones, how these factors relate to the quality of the microphone, and then use this info to dispel a few myths:

- *Price Factor 1:* What do you want in a microphone? A $250 large diaphragm condenser mic with a tube circuit, selectable polar patterns, bass roll-off, 10dB pad switch, power supply, and a shock mount should arouse suspicion in the clever shopper. For the same price, there are other mics that offer a cardioid-only pattern, with no bells and whistles. Which mic do you think uses higher quality components? Probably the simpler mic. There are no hard and fast rules, so consult trade magazines, experienced users, blogs, and Internet reviews—and trust your ears. You may wonder, how do some mics with the same features as the bells-and-whistles $250 mic wind up costing $2,000 or more? Keep reading!

- *Price Factor 2:* Where was the microphone manufactured, and where was it assembled? Since manufacturing has migrated east over the last decades, it is common to find microphones manufactured and sometimes even assembled in the Far East (especially China). This is not a watermark for quality either low or high quality, but just a fact of life. However, companies that manufacture and assemble in Germany, for example, incur much higher costs for both parts and labor. Relative newcomers to the microphone market often use Chinese-manufactured parts, but assemble, tune, and test the mics elsewhere. Microphone craftsman John Peluso (Peluso Microphones) is a perfect example, creating top-quality mics at affordable prices with this business model. Mics of the quality Mr. Peluso builds had cost several times more than the price he can now offer—an amazing value for the small studio.

- *Price Factor 3:* How much do really great parts cost? This is where quality can start to get expensive; a carefully tensioned, gold-sputtered diaphragm may cost several hundred dollars. Diaphragms based on vintage designs (like the famous Neumann M7 capsule) still require delicate handwork, and can easily cost up to a thousand dollars . . . (cringe!). Hand-wound transformers, carefully constructed head-baskets, point-to-point wiring with high-tolerance components, vintage tubes, and power supplies that supply selectable polar patterns drive prices up even further. Add all these up, and you begin to understand what a great mic costs in parts alone. For this reason, some companies offer stripped-down versions of their flagship mics, allowing budget-bound users to get a taste of the good stuff.

- *Price Factor 4:* How much does the company spend on advertisement? If you don't have a lot of experience, or the opportunity to hear a variety of mics in action, you are often at the mercy of what you read in trade magazines and on the Internet. Consider two manufacturers of similar microphones. If one company depends on word-of-mouth advertising, and the other takes out full-page ads in MIX magazine, which mic will have a higher retail price? You got it—the full-page ads are expensive, and the consumer pays for them. Does this affect the quality of the microphone? Not directly, of course, but those who advertise heavily should take care to keep quality up, as well.

When shopping around for microphones, consider these points, and find a retailer who allows you to test the mics. At the very least, they should allow you to trade back for store credit if the mic you try doesn't fit your needs. Use your ears, and remember that although quality comes at a price, at a certain point the law of diminishing returns kicks in—a $1,000 vocal mic will sound a lot better than a $150 mic, but a

$3,000 top-shelf model is not going to sound three times better than the $1,000 mic. As the quality of the parts and the care of the labor increase, you will get certain very small increases in the quality. This is often that extra edge that the pros are looking for, but it may very well be lost in the imperfect home environment. In fact, a mic that picks up the hum of your fridge behind two closed doors may be exactly what you *don't* want in your home studio!

Also, keep in mind that many classic recordings—ones that have defined the recorded tones we still want to hear today—were made with simple microphones that are still available and inexpensive today. The performance that these mics captured was the magic, not the gear itself. Beware the siren song of marketing!

Electret Condenser Microphones

Electret condenser microphones operate on a similar principal (capacitance), but use a permanently charged diaphragm backplate element, allowing for a very small final product. The little microphones in modern telephones, cell phones, laptops, and PCs are all electret condensers, making them the most widespread mic in the world, albeit not the most common for recording.

There are, however, many electret condenser mics that find their way into the studio. The company Brüel & Kjær makes a line of reference microphones using electret condenser technology, and their performance specifications include incredibly flat frequency response and fast transient response. AKG's inexpensive and rather widespread C1000s is an electret condenser, and Sony's ECM line of very cheap stereo electret microphones can make some surprisingly high-quality recordings on a shoestring budget. You'll learn a trick or two using that Sony mic in Chapter 5.

Piezoelectric Microphones/Pickups

The less common *piezoelectric microphones* have made their mark in the recording world, and not quite in the way that ribbon, dynamic, and condenser mics are used. There are existing designs in which a piezoelectric crystal is used in a microphone for picking up sound that is traveling through the air, but more commonly, these crystal mics are used as guitar pickups or drum triggers (to trigger a MIDI note for electronic drums). Lavalier mics—the little ones that clip onto a shirt for use in TV—sometimes use a piezo design, although the sound vibrations in this case are still picked up via a diaphragm connected to the crystal. Their small size, not their sound quality, makes them useful in this case.

The sound of an Ovation guitar plugged into a P.A. or amplifier is a perfect example of the sound of a piezo pickup. The electrical signal is created by a mechanical coupling of the sound source with the crystal pickup, such as in the now-widespread bridge saddle pickups. They can be made extremely small and lightweight, and therefore very convenient for live sound use. Although the sound quality may leave something to be desired, the widespread use of these pickups has put that particular sound on the map. Alice in Chains' "Jar of Flies" album (the song "No Excuses" was a big radio hit) prominently features this sound on the acoustic guitars, as does just about anything by Dave Matthews. Let your ears decide.

Drum triggers are a natural for piezo design; they can be made small and light, and directly connected to a drum, so that a drum hit also provides an electrical signal output. This signal can be used to trigger a sample which is then blended into (or even replaces) the acoustic drum sound. In this case, you likely never hear the piezoelectric signal, it is simply used as a trigger to play back a stored sample. Electronic drum kits may also use a piezoelectric element for each drum piece, thereby triggering the different sounds that make up the kit. In any case, this technology provides a unique way to create a signal via direct physical contact.

Understanding Microphones

Whew! That's a lot of technical stuff you just worked through. The good news is you can now get into some more practical examples as you learn about the qualities of microphones in general.

The basic design of a microphone affects certain aspects of its overall tone, but there are other characteristics to microphones which vary greatly, and which may not be rooted in its design. Here's a short breakdown, a sort of tonal checklist as to how a microphone behaves in the presence of sound. This checklist can help give you an idea how the mic will "shape" or "color" the sound source:

- Frequency response
- Diaphragm size or ribbon element size
- Polar pattern
- Transient response
- The microphone's sensitivity to sound pressure, and the point at which it overloads
- Additional qualities of the design, such as vacuum tubes, FET, and transformers

 So, recording the same instrument with different microphones can provide you with some very different sounding recordings. In Chapter 3 of the CD-ROM, you can listen to an example acoustic guitar recording. The same performance was recorded simultaneously with a Shure SM-57, a Golden Age R-1 ribbon microphone, and a Peluso 22251 Tube Condenser microphone.

Have a listen to each one, and think about your initial impression of the sound in each case.

The reason I chose these three microphones is that they all have a different basic character. The Shure 57 is a dynamic mic, the Golden Age R-1 is a vintage-style ribbon mic, and the Peluso 22251 is a large-diaphragm condenser microphone. Their frequency response, basic transducer type, polar pattern, and overall design are all rather different, and the sound they produce from the same performance is distinct.

Let's learn a bit more about these characteristics, and how they affect the sound of the microphone, and thereby your choice of which mic to use for a given sound source.

Frequency Response

You are probably familiar with a dog whistle, which creates a tone too high for human ears to hear, but still falls within a dog's hearing range. The dog whistle is a good analogy for understanding frequency response and how it applies to recording.

Let's imagine that your ear is the microphone, and the dog whistle is the sound source. In this case, your ear doesn't have the high-frequency response necessary to hear that high dog whistle tone. Microphones behave this way as well, a characteristic called *frequency response*. This describes the spectrum of sound that the mic can pick up, and how sensitive it is to particular frequencies.

Our ears can pick up sounds with frequencies from 20 cycles per second to 20,000 cycles per second (20Hz to 20kHz). The unit for frequency, hertz (abbreviated Hz), was established in honor of scientist Heinrich Hertz for his contributions to the study of electromagnetics. You'll hear this unit of measurement referred to in audio engineering more often than any other:

- "Can you cut a little 360Hz out of that kick drum?"

- "Those cymbals are a little dull. Try boosting some high shelving at 12kHz."

- "There was a rumble coming through the mic, so I cut everything below 50Hz."

All of these statements refer to the use of an equalizer (EQ); one of the most important tools. EQ allows you to further manipulate the frequencies in a sound signal, usually to correct uneven frequency ranges in a sound, but also for unique effects. (Chapter 6, "Signal Processing Toys: EQ," is dedicated to EQ tips and tricks.)

So, you could imagine that an optimal microphone would have a perfectly even response from 20Hz to 20kHz, so that it would fit to the human ear's natural range. There are some mics available that can just about pull that off, but they aren't as popular in studio recording as you might expect. When recording an orchestra in a concert hall with excellent acoustics, a pair of microphones set up to mimic a person sitting there listening would give you an excellent natural representation of the orchestra in the concert hall. This is a great way to approach such a recording, but likely doesn't apply much to the art of studio recording.

Recording classical or even traditional/folk music may certainly require you to capture an accurate "picture" of the performers, but because this book focuses on the home studio, the situation is different. You'll at the very least have to cope with some imperfections in the recording environment, if not also with less-than-perfect instruments.

The truth is, studio recording often involves a series of manipulations, meant to flatter the sound source and bring out particular qualities of instruments for a larger-than-life effect. It corrects for imbalances in a particular sound, and aims to fit all the instruments into the mix; leaving a certain amount of aural space for every element. Starting with the late 60's recordings of Phil Spector, Joe Meek, Motown Records, The Beach Boys, The Beatles, The Rolling Stones, and other early pop/rock pioneers of studio manipulation, the role of the studio changed from a place to simply lay down the music that artists played onto vinyl or tape into an instrument itself—a place to experiment with sound and discover ways to make one's music stand out in the crowd.

Rock and other popular forms of music aren't generally performed in meticulously designed concert halls or listened to in carefully balanced hi-fi sitting rooms. This music is meant to be heard on boom boxes, car stereos, jukeboxes, and headphones; delivered via radio, webcasts, MP3s, and other imperfect formats. Astute engineers discovered long ago that clear, accurate recordings don't always translate well to these formats. By tweaking, twisting, and torturing the sounds they record, an engineer/producer can help the sounds jump out of the speakers and grab modern listeners, helping the music to get the attention the artists crave.

The first step in this philosophy is to pick a microphone that will bend the sound in the direction you want it to go. Let's consider the studio must-have Shure SM-57 microphone. This mic's frequency response looks approximately like the one shown in Figure 3.4.

What you see in Figure 3.4 is that the Shure SM-57 does not respond evenly from 20Hz to 20kHz. Does this make it a lousy microphone? Not at all. In fact, it so happens that the way it responds offers a few advantages. Let's learn how to read this chart, and how this info can help you make better recordings.

First, notice that frequencies of 50Hz and below are picked up 10dB lower than frequencies in the middle range. This ramps up to become more or less even around 100Hz.

It so happens that many consumer radios, boom boxes, and the like don't reproduce frequencies below 60–80Hz (or so) very well. Lower frequencies also require more

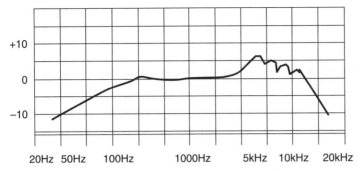

Figure 3.4 The Shure SM-57 has a characteristic frequency response that was originally designed to add an extra boost to the midrange frequencies that enhance the intelligibility of speech in a public speaking situation. It so happens that this frequency curve works well for many applications in the recording studio, from snare drum, toms, and percussion, to guitar amps, and even the occasional vocal recording. Some engineers dislike the 57 for the same reason, calling it harsh or thin sounding.

energy to reproduce, which saps energy from the amplifier. So, by not capturing these frequencies, the mic is filtering out less useful sound, allowing more energy to be put into the range you hear on that boom box. The result is a more fitting recording for the lo-fi situation, a sound that has more power in an effective hearing range.

Naturally, this attribute would be a weak point if the source you are recording has an extended low-frequency response, such as low percussion or drums, bass, or a particularly low-ranged vocal.

Next, have a look in Figure 3.4 at the bump in the frequency response around 5–6kHz. Human ears are particularly sensitive to sounds in the 1kHz to 10kHz range, because this is where much of the detail in spoken language sits. So, a bump in the 5–6kHz range will make a sound seem louder; it will cut through the mix, that is, bring the sound forward relative to other sounds.

You might not always want sounds to cut through in this way, because the recording can become harsh sounding. So, you must choose the right mic for the job; try to tailor fit the mic's character to a sound source for best results. For example, mic'ing a snare drum with a Shure SM-57 (a popular mic for this application) applies this natural low-frequency roll-off and 5–6kHz boost to the snare drum sound. On the right snare drum, this can be a match made in heaven—the snare drum gains the right edge to stand out in a pop/rock mix. Using your ears is important—if the snare drum sounds harsh or lacks low end, an alternative mic choice would be the first remedy.

Looking one last time at the frequency response graph shown in Figure 3.4, you'll notice that the SM-57's response drops off quickly at 15kHz—the high end of the audio spectrum. As a dynamic mic, the 57 uses a relatively weighty transducer design, making it less sensitive to (naturally lower-energy) high-frequency sounds. For this reason, it would be unwise to depend on a Shure SM-57 to clearly record drum cymbals, the clarity of plucked strings, or other sounds that have particular high-frequency detail. For the crunch of an electric guitar or the pop of a snare drum, however, this mic is often a good match.

Tip: As an aside, some engineers really hate Shure SM-57s, saying that they sound harsh or one-dimensional. Opinions vary and everyone is entitled to one, so give this mic a try on lots of different sources and build an educated opinion for yourself. I've captured scratch vocals on a Shure SM-57 before that I was never able to improve upon with $2,000 condenser microphones. For its under $100 price tag, keep at least one around!

Now you know how to read a frequency response chart, and understand how it affects the output of the mic relative to the input (sound source). Think about how different frequency responses may fit to different sources.

To what might you match a microphone with a ruler-flat response?

How about a "smiley face" response (lots of sensitivity to bass and high treble, but less in the mids)? How about a "frown-y-face" response (accentuated midrange response)?

More to come on that in Chapter 5; just some food for thought right now. Let's finish learning about mics in general.

Diaphragm Size

The size of a mic's diaphragm affects its sound in a couple of ways. I talk about condenser microphones here, as they generally fall into two categories—small diaphragm (one-half inch or less in diameter) and large diaphragm (generally one inch in diameter).

A larger diaphragm naturally has more weight, and is therefore slower to respond to sound vibrations. A diaphragm also has, as does every object, a resonant frequency. Larger objects of the same material have *lower* resonant frequencies than smaller ones, so large diaphragm mics have a lower resonant frequency, which means they have a generally warmer sound.

On the other side, small diaphragm condenser mics respond faster to transients, creating a more accurate or detailed sound in comparison to their big brothers. Additionally, sound arriving at an angle to the diaphragm tends to be reproduced better by small diaphragm mics. Sound arriving from the side of a large diaphragm may cause *phase problems* if the peak of the sound moves to one side of the diaphragm while the trough of the wave hits the other. Smaller diaphragms have less of a problem with this effect.

Polar Pattern

The direction in which a microphone 'listens' for sound is known as its polar pattern. *Polar diagrams,* as shown in Figure 3.5, show you exactly how this works. They're easy to read; just imagine that the north pole of the diagram is the direction the mic is facing, and the south pole is the back of the microphone. This leaves you with east and west, which are simply the sides of the microphones.

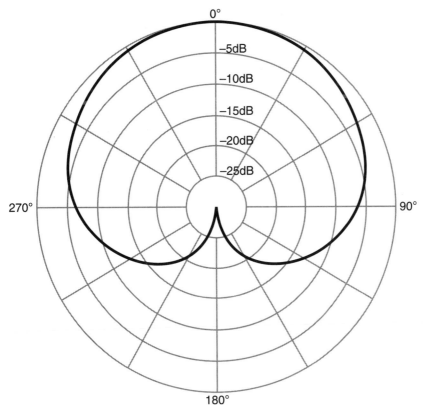

Figure 3.5   A graph of cardioid polar pattern's sensitivity to incoming sound. Sound arriving at the front of the mic is picked up more effectively than sound arriving at the back. This effect is helpful in reducing reflections and "bleed" from other sources.

So imagine that you are singing into a microphone, and there is some sort of sound also coming from behind the microphone. Perhaps it's your thousands of screaming fans. You love your fans, but you're the one making the noise to be amplified here, not them. In this case, a microphone with the polar pattern shown in Figure 3.5 is the right choice.

The name cardioid was chosen due to the heart shape of the polar pattern diagram, as in cardiac, a word with Greek roots.

This polar pattern is one of the most useful in both live sound and recording, because it rejects sound coming from the back of the microphone, while still providing a full sound from the source at the front. This makes it a great choice for any spot microphones in a situation where nearby instruments may also be blasting away (snare drum, toms, or guitar amplifiers) and you want to reduce bleed. *Bleed* is what engineers call unwanted sound that gets picked up along with the sound source you intend to record. The "heart" reduces bleed? Go figure...

Typical problem sounds are loud cymbal crashes, loud guitar amplifiers, loud singers... you get the point. Anything loud and near your sound source. With cardioid pattern mics and careful positioning, you can often reduce bleed to acceptable levels while still allowing the musicians to perform in the same space. The Shure SM-57 has a cardioid polar pattern.

Going one step further, there are *hyper-cardioid* polar patterns, which reject even more sound from the back and sides of the microphone, generally at the sacrifice of some roundness of tone. See Figure 3.6.

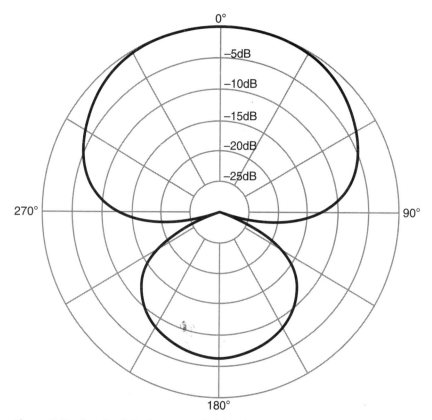

Figure 3.6   Graph of the hyper-cardioid polar pattern.

The hyper-cardioid polar pattern is generally used for live sound situations where bleed is a more serious problem, leading to incoherent audio channels and feedback. The drawback to the hyper-cardioid pattern also lies in its narrow pickup pattern, which rejects some good along with the bad. You should experiment with hyper-cardioid mics to get to know their characteristics, but in general, avoid them for studio use.

More useful in controlled studio situations where you want to pick up room tone along with the sound source is the *omnidirectional polar pattern,* as illustrated in Figure 3.7.

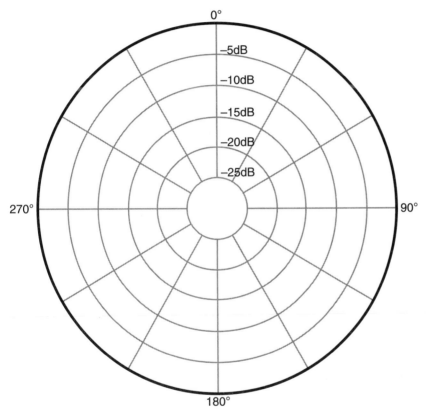

Figure 3.7   Graph of the omnidirectional polar pattern.

In this case, the microphone picks up sound equally, regardless of the direction the sound is coming from. This is obviously not the right choice for a single instrument in a band setting, as it will pick up far too much bleed from other instruments. What it can do very well, however, is pick up an overall picture of the band, which will give the final mix a sense of space and ambience.

The omnidirectional pattern is also an excellent choice for recording a single instrument in a room with good natural acoustics. When recording drums, an "omni" mic (or two) is a must-have. Piano, guitar, percussion, and traditional acoustic instruments often sound great when recorded with an omnidirectional mic due to the balanced, natural sound it can provide. Vocal recordings made with an omni mic will have a pronounced sense of the space you are recording in, for better or for worse. For this reason, the standard pattern choice for a vocal mic is cardioid unless you are seeking a particular effect.

The last major pattern, which is typical of ribbon microphones, is the *figure-eight polar pattern,* as shown in Figure 3.8.

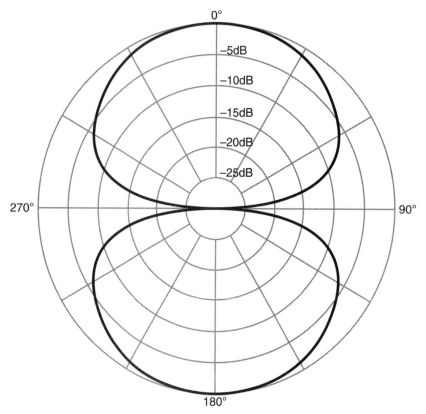

Figure 3.8 Graph of the figure-eight polar pattern.

This pattern rejects sounds coming from the sides of the mic. Ribbon mics tend to have this polar pattern due to their basic design. Figure-eight pattern mics tends to have a natural, open sound similar to the omni pattern. Cardioid and hyper-cardioid require a more complicated design to create this polar pattern, resulting in some (if minimal) distortion of the incoming sound. Some purists like omni and figure-eight mics for this reason.

The figure-eight pattern also allows you to get up to some clever trickery as you get into multi-mic'ing sounds and stereo techniques.

Many large-diaphragm condenser microphones offer a *selectable polar pattern*, allowing you to change the polar pattern at the flip of a switch. You can then easily compare the sound that different patterns provide, and choose the right one for your current situation. Some fancier models even provide in-between settings, with polar patterns between hyper-cardioid and cardioid, cardioid and omnidirectional, and so on. I've had great results with some of these settings, especially when varying between omni and cardioid and fine-tuning the room mics on a drum kit.

Transient Response

The *transient* of a sound is the leading edge "attack" of the sound, and is very important in defining the quality of the sound. The hammer strike on a piano string, the rap of a stick on a drum, and the rasp of a pick on a guitar string are all quite familiar to our ears. Many listeners consider a good recording to be one in which this delicate, fast bit of audio information is clearly reproduced.

The weight of the transducer element in the microphone affects the transient response of the mic. A heavier transducer (such as a large, plastic diaphragm) has more inertia, and takes longer to begin moving in sympathy with the sound energy than a lighter transducer (such as a four-micron-thick aluminum ribbon).

A fast transient response adds a sense of realism to the recorded sound, but isn't inherently "better sounding." A mic with a slower transient response, like a dynamic microphone, may color the sound in a pleasant way, sounding "softer" or perhaps even more like a "recording" instead of a realistically "captured" sound. Naturally, lighter transducers tend to be more fragile and prone to distortion, which is a consideration when mic'ing certain instruments.

If you want to think in very general terms (there are always exceptions), the general transient response of mic styles from fastest to slowest is ribbon mics, small diaphragm condensers, large diaphragm condensers, and then dynamic mics.

As noted, there are exceptions, and some dynamic mics have excellent transient response. You'll need to get to know your own mics over time to help you pick the right mic for the sound source. If you are planning on purchasing a few mics, it's a good idea to have a broad range—maybe a few dynamic mics, a large diaphragm condenser, a pair of small diaphragm condensers, and if budget allows, a ribbon mic as well. With this spread of mic styles, you'll be well armed to handle anything from recording vocals to a whole drum kit.

Sensitivity to Sound Pressure

Most modern mics can handle fairly loud sources without problems, but you may run into situations where a hard-hitting drummer or a particularly powerful vocalist pushes the microphone itself "into the red," and you start getting some nasty distortion. Condenser mics tend to be the most sensitive here, because their internal electronics provide a relatively strong output, and this can overload elements in the signal path at high SPLs (sound pressure levels). Transformer-equipped mics tend to distort more readily as the transformer "saturates" (reaches its dynamic limit), although when not exaggerated, this mild overloading can add a pleasing warm character to the sound. Transformer-less mics can just about handle dynamite blasts without trouble, making them a good choice for drum recordings.

Some mic models have a –10dB pad switch for loud situations, which reduces the mic's sensitivity and saves the day. If you're using a mic that doesn't have a pad, and you hear otherwise inexplicable distortion during loud performances, you may need to switch mics. There's nothing worse than kick drum hits that start to sound like elephant farts in the middle of the chorus...except finding out that your expensive condenser microphone's diaphragm has collapsed, and it is now the prettiest doorstop in the studio!

Despite the otherwise delicate and wind-sensitive element inside, ribbon mics can generally handle mountain-shaking SPLs before distorting. Dynamic mics are the runners-up, and I don't think I've ever heard a Shure 57 or a Sennheiser 421 pushed into self-distortion, so they're a safe bet for mic'ing up that smoldering stack of Marshall JCM-800s glowing with anticipation in the garage.

Additional Mic Design Qualities

The design of a microphone's internal electronics influences its overall tonal character, especially when it comes to the way a mic behaves when things get loud. Transformer-less microphones tend to fall on the more "accurate sounding" side of the spectrum, whereas FET and tube electronics might add a particular character to their microphones. Note that this statement is qualified with the word might—not all that glitters is gold, nor is all that glows really "tube-y" sounding.

Tip: *Tube electronics* has been, by far, the biggest hype phrase in the last decade, with every type of gear from microphones to compressors to EQs to D.I. boxes (*direct injection* boxes, which accept a variety of inputs for recording bass, guitars, and keyboards direct to tape) being loaded up with cheap tubes so that gear manufacturers can sell more toys. The truth is, a vacuum tube is just one element in an amplifying circuit, and it may have little or no effect on the sound quality. In fact, proper power supplies, quality capacitors, and phase-coherent audio transformers contribute much more to a seriously professional sound than slapping a tube into the circuit. The problem is, quality transformers may cost hundreds of dollars a piece; a cheap tube may cost a few bucks. Hype, ladies and gentlemen—you have been warned.

That said, use your ears, not the glossy pages of the latest *Audio Astronauts Review* magazine to pick your gear. A good music shop will let you audition a mic if you are serious about purchasing some quality gear. Some now refuse returns due to "hygienic issues," but if you discuss your intentions beforehand, an exchange should be no problem for a serious audio dealer. In this way, you can audition a mic in your own setting and see if it has the mojo you want.

Transformers are another significant element in microphone design. A transformer serves several purposes in audio design:

- A transformer physically isolates one circuit from another while allowing audio (an alternating current signal) to pass. Without getting overly technical, this helps designers control unwanted noise.

- Transformers can step up the voltage of a signal, allowing for a stronger microphone output. Up to 25dB of gain can be achieved with a transformer, allowing for better signal transmission through cables, and less gain needed from potentially noisy preamplifiers!

- Transformers sound pretty cool. Yep, that's another reason to use 'em! Although a good quality audio transformer can add several hundred dollars to the cost of a microphone, their natural characteristics "color" the microphone signal in a pleasing way. You can think of a transformer making the sound of a mic "thicker," "smoother," and more recorded sounding. Again, this is just a rule of thumb. Makes and models abound, and there is essentially no price ceiling. You can spend almost $4,000 on a Blue Bottle, $2,000 on a RODE Classic, $1,000 on a Mojave MA-200, or just $200 on an Apex 460, and in any case have a mic that sounds pretty darned spanky.

Mics using FET design can also color the mic sound if that is the designer's intent. Examples include the Neumann U47 FET ($5,000 or more), MXL V-6 ($300), Apex 480 ($225), and the Mojave Audio MA-201 ($700). I recommend searching the Internet for reviews and shootouts among different mics. It often happens that when the "voodoo factor" of the mic's manufacturer name is removed, less expensive mics prove to be just as good-sounding on many sources.

Transformer-less FET mics provide cost-saving alternatives to many manufacturers tube-and-transformer flagship microphone models with their inherent bulky power supplies and extra cable-spaghetti. The success of Neumann's TLM (transformer-less microphone) transistor mics seems to have led the way for many other manufacturers, who now offer similar designs. Transformer-less FET mics that have recently made their way into the mic lockers of home and professional studios alike include the Neumann

TLM-103 ($900), AKG AT4033 ($400), and the RODE NT1-A (just $200, and a very popular and useful mic).

Picking the Best Mics for the Job

Food for thought: As an engineer, choosing your microphones is like a painter choosing canvas, paints, and brushes. These are the tools of your trade, and quality does demand a price. However, just because a brand name demands more money, that doesn't always mean that all the money spent goes into the quality. Many companies spend a small fortune on advertising, so part of the expense is linked to their brand name. On the upside, Neumann microphones (rather high-profile) tend to maintain their resale value. Something to consider. However, here are a few tips for building up your mic locker:

- Be sure that you have plenty of "workhorse" mics. No, I don't work for Shure, but I must say that a few SM-57s are a must-have for any studio. There are imitators out there, but beware of the cheap Chinese-manufactured copies, as they often suffer from poor components, and lax quality control, both of which lead to lousy sound quality. Investment for your studio: $300 or so.

- Ribbon mics are making a comeback, and you shouldn't miss out on their smooth, balanced sound, which can take the harsh edge off of digital recordings. Top-notch mics from Royer, AEA, and Coles price in at about $1,000 each, and have the tone to prove it, but lower-cost ribbon mics are cheap and very useful as well. The Nady RSM-2, Apex 210, CAD Trion 7000, Cascade FAT HEAD, and Golden Age R1 (among many, many others) cost somewhere in the $150–$250 range, and make a fine addition to the budget-conscious engineer's cabinet.

- A pair of small-diaphragm condenser mics shouldn't be missing from your toolbox, either. They'll come in handy when recording stereo drum overheads (or as room mics), acoustic guitar, piano, background vocals, location recordings, and in any situation where you want an accurate stereo recording. Again, prices range from $2,000/mic to $100/mic. Less expensive mics often sound thin or brittle, whereas top-notch mics sound utterly balanced, smooth, and realistic. Schoepps and Neumann cover the top of the market ($1,000 per mic range) with attendant quality; Schoepps is generally known for accuracy, and Neumann for their particular "larger-than-life" sound. The AKG C-451, Shure SM-81, Peluso CEMC-6, Oktava 012, and Mojave MA-100 sit in the mid-market range, offering excellent mics for the price—around $500/mic. The RODE NT5 and MXL 603 are good examples of the $300–$400/pair options available. If possible, try different mics before you buy. Because these mics are generally not used for close vocals, your local dealer may let you audition some for a deposit. Try them on drum overheads and acoustic guitar, listening carefully for harsh or unbalanced qualities; as with all things in your studio, the room and instruments may affect your mic choice more than brand reputation or web-based reviews!

■ You will probably need one large diaphragm condenser mic, and this mic will seriously affect your sound; the sound of vocal recordings is what most people listen to first when checking out a studio's work. As mentioned, you can easily break the bank with a large diaphragm tube/transformer/multi-pattern microphone. No need to panic though; the market is so flooded with quality large diaphragm mics, that you'll have no problem here. If you want a solid start, grab yourself a RODE NT1-A, Audio Technica AT-2020, Shure KSM-27, or KSM-32 (to offer a few reliable inexpensive models) and you'll have a mic that you will want to keep even as you expand your mic collection. There are a few cheap tube/transformer mics that offer a more "colored" sound for just a few hundred bucks, but I recommend adding one of these as a second choice, because the colored sound may not work as reliably on different sources. The Apex 460 is a great choice ($200), as are the slightly more expensive RODE NTK, RODE K2, Groove Tubes 6TM, MXL V69, and ADK TC. Prices go up and up from there, with Peluso, Mojave Audio, Charter Oak, Neumann, Brauner, Blue Audio, Soundelux, Lawson, Manley, and many other boutique mic manufacturers offering high-quality mics at high-quality prices!

Overall, don't give in to the price game, or worse yet, the "gear acquisition bug." The "gear bug" causes you to spend more time checking out gear than using what you have—believe me, I've had it, and it's a big waste of time (and money). The truth is, all these toys don't sound so radically different, and you'll wind up just reselling a lot of them.

All in all, if you follow the most price-conscious suggestions, you can have a mic locker with all the toys you need to start playing games for less than $1,500. It might break down like this: Four Shure SM-57s ($300), a pair of RODE NT5s ($400), a Nady RSM-2 ($200), a Shure KSM-27 ($200), and an Apex 460 ($200). That's just $1,300, and you'll be able to mic up anything that's thrown your way. If you use local pawnshops, want-ads, eBay, and Craigslist, I'd be willing to bet you can keep the total cost under $1,000. After you set up the studio in the next chapter, you'll learn how to mic up all the instruments that may make an appearance there.

Using Multiple Microphones: Understanding Phase

Often misunderstood or just ignored, the concept of *phase* is simple to understand, and can help you greatly in your quest for better recordings. The first thing you need to understand is that phase is *relative*. It can be simply put that phase effects occur because a sound arrives at different times at different points. If the time is long enough, you may hear it as a delay or echo-type effect, but when the time is so small as to be within one complete cycle of a sound, then you're down to the very small time frame wherein you consider phase.

You may already know that adding two waveforms will give you a new one as a result. Well, any time you have more than one microphone on a single sound source, you should consider how these two similar signals will be combined. Phase really only applies to two signals that are rather similar in frequency content, as they are similar enough to cancel one another out. If both signals arrive at some point (like a microphone) simultaneously, they are in phase. If not, they are out of phase.

Consider two situations; in phase and out of phase.

Two signals that are in phase with one another will combine positively, and create a louder signal. Pretty simple. Think of two waves in the ocean piling up to create a bigger wave. They are *in phase*. See Figure 3.9.

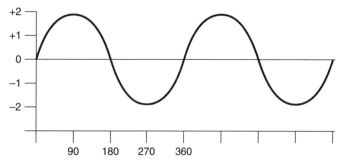

Figure 3.9   A representation of the elements of a wave signal. The amplitude is represented on the X-axis (vertical), the Y-axis (horizontal) represents time, and is measured in degrees as well. Imagine that the degrees measurement (Y-axis) shows you when one full cycle or "rotation" has occurred. Waves are cyclical, just like rotation.

If two signals are completely out of phase, they cancel each other out; the positive energy of one adds to the negative of the other, and you get zero. These signals are *out of phase*.

Those are the two simplest situations, but the signals could also be slightly out of time, in between in and out of phase. This is a bit more complex to deal with, and presents a particular challenge in audio engineering. To describe how much out of phase the signals are relative to one another, musicians use the term *phase shift*. Phase is measured in degrees, a full cycle of a waveform being 360 degrees (just like a circle). See Figure 3.10.

Tip: Unless you're dealing with synthesized waveforms, you'll rarely find two signals in the real world of recording that are perfectly similar, so the images are to be taken with a grain of salt. In very dissimilar signals, their phase relationship is not even an issue. Many engineers use the "three-to-one rule" to avoid phase problems when

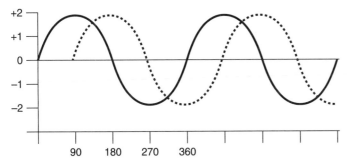

Figure 3.10 The signal shown as a thick line is the reference; the dashed line shows a signal that is out of phase by 90 degrees. Because this example deals with very small fractions of a second (less than a millisecond), you do not hear two separate signals, but rather the sum of the two, and this phase difference will audibly affect the sound of the signal.

> using multiple mics. To do this, measure the distance from the closest mic to the sound source. Then, make sure the second mic is at least three times this distance from the source, and in most cases, the signals will sum without phase problems.

In the real world, as you may have also experienced, nothing is perfect. When two audio signals coming from one source are summed, they are rarely going to add or cancel perfectly, so the result will be a blend of the two sounds, and not always for the best. In this case, certain frequencies may cancel, and others pile up to become louder, leaving you with a hollow or imbalanced sound. This typically happens when a very close reflection of a sound meets the direct sound at a microphone.

A typical example is recording upright bass or bass guitar. Quite often, you'll take both a direct signal and a microphone signal from the bass and blend them in the mix for a combination of clarity from the direct line and tone from the mic'ed amplifier or instrument. A simple and useful tool for checking to be sure you have a solid bass tone is the *polarity switch* (sometimes inaccurately labeled by manufacturers as the *phase switch*). Many EQ plug-ins have this feature, so if your preamp doesn't have one, don't panic—here's what you do.

Pan the direct sound and the mic'ed sound to the middle of the stereo field. Make sure that they are playing back at pretty much the same volume (big volume differences could hide the problem). While listening to the tracks playback, flip the − polarity on one channel. You only need to flip it on one; if you flip both, you'll negate what you're doing!

Do you hear a major difference when you flip the polarity on one track? You should. If the volume drops significantly, or they sound hollow or odd together, flip them back, and you can be confident that they were well in-phase at the start. If flipping the phase switch actually brings them into focus (that is, they initially sounded lousy), leave the switch engaged. You've found your solution—great engineering!

If you hear no change, one of two things is going on:

■   *The tone of the two tracks is so different that they aren't lined up enough to cancel in any case.* You're AOK if this is the case; just continue tracking.

■   *The tones are similar, and they are about 90 degrees (a quarter of a cycle) out of phase.* When this is the case, flipping one just throws you into the same situation, upside down, so to speak. You need to adjust something. Try moving one microphone a few feet toward or away from the source. Repeat the test, and see if it improves. With a few attempts, you'll find a sweet spot where the sounds reinforce one another and create a solid tone.

This technique should be used any time you multiple-mic an instrument. It is especially important on bass drum, toms, bass guitar, and anything with prominent low-frequency energy, but the rule applies to any multi-mic'ed situation. When bass frequencies cancel you'll wind up with certain bass notes just disappearing in the mix, making the song sound empty. Checking phase (polarity) will help you root out wimpy tones from the get-go, lay down solid tracks, and save you later mixing headaches.

4 Setting Up Your Studio

Now that you have a good understanding of how recording is supposed to work, you're ready to put together a studio. But before you start ripping into those boxes, tossing user's manuals out of the way, and dashing for the nearest electric outlet with a fistful of cables, you need to set up your studio.

The key to great recordings, as it is with great food, is to have the proper, quality ingredients. When it comes to recording music, you must keep in mind that it is recording *music* that is important, not just the *recording*. I'm sure you have some favorite music recordings in which the sound quality is not 100%, but the performance has a certain something to it that grabs you every time you hear it. It could be an old blues record, a live recording, or just a record made before the days of 24-bit 48kHz digital such-and-such. In this way, you can determine that the key aspects to a great recording are as follows, in descending order of importance:

- A great song, and an inspired performance
- Good-sounding sources (well-tuned drums, good guitar tone, a singer who doesn't sound like Fran Drescher)
- A decent-sounding room
- Good engineering technique (mic placement, mic choice, processing, and mixing), including a good monitoring situation
- Quality gear that is well connected and maintained

"What? Good gear comes last!? Why did I spend all that money? I think I hate this book."

Hang on! If there's one thing I've learned by looking back at some of the recording greats, it's that the gear was never the most important link in the chain. It was always the concept, inspiration, performance, and willingness to experiment. Naturally, some bits

57

of gear will sound better than others, but the quality of even hobbyist gear today is quite good. So that's actually the least of your worries.

Think about it for a moment. Imagine that you're not going to record, but just listen. What do you want to hear? A tone-deaf grade-school dropout playing an out-of-tune violin in a gymnasium? Of course not. First of all, you want to hear a great song, preferably performed by someone with a decent instrument and voice. I'm going to assume that your talent is ready and waiting, instruments in hand, for the studio to be set up.

Knowing the Home Studio Pitfalls

This brings you to the room in which you will record. Because this book is about home studios, you'll probably be converting a spare bedroom, finished basement, or garage space into your studio. These rooms bring a few basic problems along with them, which you'll learn how to fix with a little elbow grease and a few trips to the hardware store.

The basic problems with using your home as a studio space all have to do with acoustics:

■ *Sound coming in and going out.* The neighbor's lawnmower sneaks onto your banjo tracks. The 47th take of a bagpipe overdub finally pushes your roommate over the edge...

■ *Room resonances.* This is the effect you may have heard when listening to music in a sparsely furnished room; some notes seem to well up and become louder, especially bass notes.

■ *Room reflections/echoes.* Clap your hands in a room with hard, parallel surfaces. Hallways are a great example. That fluttering, echoing sound is a series of short, fast echoes that are pretty lousy sounding for recordings. *Comb filtering* is the effect that these pinging echoes have on the frequency response of a room. Imagine a frequency response graph that looks like the teeth of a comb, and without needing a Ph.D. in acoustics, you'll understand that this ain't a good thing.

This chapter focuses on how you can remedy these problems without breaking the bank. For this example, I assume you are using an existing room. If you are going to build the entire studio from scratch (floor, walls, and all), I recommend you find a dedicated book on the topic of studio construction.

You can do the following adjustments in a modular sort of way, adding elements as time and budget allows. It's also nice to keep these elements moveable in case you need to move to another space at some point. Every step you add will improve the sound of your recordings, so if you have even a hundred bucks more in your studio budget, consider investing in your studio's acoustics!

Before continuing, I must acknowledge the help and inspiration gleaned from my now dog-eared copy of F. Alton Everest's *Master Handbook of Acoustics*. If small studio acoustics interests you and you want to learn more about the topic, that handbook should be your very next investment.

Before You Start: The Don'ts of Sound Treatment

To soundproof a room is far beyond the scope of this book. If you feel the need to stand outside your studio and hear *nothing* while Ozzfest rehearsals are happening inside, you'll need a team of acousticians and architects and the budget to match. This chapter instead talks about sound *treatment.* You're going to control and reduce sound transmission to create a quieter and more even-sounding studio space.

The Internet is a notoriously lousy source for advice on sound treatment/soundproofing, among other things. Trial-and-error is also a bad idea, because many approaches you may think would help often don't, and in the worst case may actually make the room sound worse. So before you run out to the local overstock store, make sure that you read this don't list:

- *Don't use carpeting as acoustic treatment.* In fact, you may not even want to carpet the floor in your studio. Keep some throw rugs handy in case of cold feet, but the reflective floor won't usually be a problem. The worst idea *ever* in amateur studio design is to carpet the walls. I worked in a project studio in the Bronx once where the dude carpeted the walls, floor, and—get this—the ceiling! The room sounded so awful that I had to quit. There's no way to get a good sound when all the high frequencies are absorbed by carpeting, and the lows bounce around like a big, muddy echo chamber.

- *Don't use foam rubber mattresses or mattress liners in place of acoustic foam.* Foam rubber is flat out dangerous in a fire, and will also eventually turn crumbly, dusty, and just plain ugly. The brand name Auralex foam is rather expensive, but there are companies now making the same quality foam at a much lower price, and most importantly; it is fireproof.

- *Don't use egg cartons as sound diffusers.* They are not effective, and are probably not all that fireproof, either.

- *Don't get involved in a big construction project such as a floated floor or such, regardless of what you read about it.* Financially, it is far too much for the small home project, and if done wrong can lead to bigger problems.

- *If you can help it, don't choose a square room for the studio or control room.* Square rooms have the most difficult-to-tame acoustics, due to the strong resonances that occur between the walls, floor, and ceiling. Rectangular is better, and if there happens to be an odd-shaped room at your disposal; that's even better.

When it comes down to it, the human ear perceives sounds that last longer as being louder. By keeping sound from transmitting freely into or out of the studio room, controlling room resonances, and creating some diffusion, you'll improve your home studio to a very good functional point without breaking the bank.

Step 1: Sound Isolation

This is where you keep outside sounds out, and inside sounds in. I find, especially when singing a song for the first time, that I really don't need to broadcast it through the whole house. Nor do I want to be interrupted by barking dogs, sirens, or the neighbor's grill party.

The key to this problem: Air is the best way to transmit sound, and mass is the best way to stop it.

This means that even a keyhole in the door will allow unacceptable levels of sound in and out of your studio. The added mass of sound absorbers and the mechanically de-coupled separations between the studio and the outside world (like double doors and double glass windows) will be your best friends when building your private, creative bubble.

It works like this. Air readily transmits sound energy, solid objects reflect or absorb it. So if you aim a speaker at a door (as will be the case in your studio control room), the door will reflect and absorb some of this energy, and any air space at all—such as the keyhole or the space between door and floor—will allow sound to pass. If you seal the door with weather stripping, then you help stop the air-based transmission of sound.

What will still be a problem is the door mechanically transmitting some of the absorbed sound energy. The good news is that this transmission is pretty inefficient. If you add a second door, you can significantly reduce this problem without having to resort to using a massive, vault-like single door. The sound would have to be transmitted by the first door into the airspace, absorbed by the second door, and then broadcast again into the outside world. Or vice-versa to get into the studio. Hence, the effectiveness of double doors in studios.

The same goes for windows. Double glass is okay; double insulated windows are even better. Windows in your home studio will be a compromise in most cases, especially when it comes to isolation against lower frequencies (which carry more energy). But if you can feel a draft from the window, it's time to do some home improvement—where air can sneak through, sound will get in and out!

Wall, door, and window installation for optimum sound insulation is an art unto itself, and is a topic broad enough for a dedicated manual. I highly recommend both *How to Build a Small Budget Recording Studio From Scratch: With 12 Tested Designs* by Mike Shea, and *Sound Studio Construction on a Budget* by Alton Everest, if you are looking to build from the ground up.

Risers and Platforms

Physical (contact) transmission of sound is a different consideration, and you can build some easy solutions to help your small studio without having to alter walls and floors. This is most practical for the home recordist who rents a space, and cannot alter existing walls, floors, and ceilings.

Drums, percussion, and even cranked up guitar or bass amps can create enough physical energy to transmit through walls, floors, and ceilings by way of their resting on the same surface. A step in the right direction is to build a riser on which to set the drum set or amplifier to decouple it from the floor. If you have downstairs neighbors, this is the place to start!

These are pretty simple projects, should cost less than $100, and will make your studio considerably more neighbor-compatible.

For a *small guitar- or bass-amplifier platform,* get yourself some rubber foam 4–6 inches thick, and cut a piece of plywood of the same size to fit on top of the foam. The foam will suspend this new surface above the floor, reducing physical transmission of vibrations. This is also helpful if you are creating a separate control room and tracking room and aren't able to separate the floors. A cranked guitar amp will pump low frequencies along the flooring and into the control room, blending with what you hear from the monitors. Decoupling the amp from the floor will help you monitor more accurately and make better tone decisions!

A *drum platform* can also work as a platform for larger amplifiers, although not simultaneously, for obvious reasons. Because the combined weight of a beer-bellied drummer and his 37-inch bass drum–equipped kit may be a few hundred pounds, you'll need to do a little figurin' here to develop a solid, effective platform. Be sure to adjust these rough calculations for your own situation.

I recommend using *machine isolation pads,* sometimes also called "vibration pads" or "vibration reduction pads." These are easy to find, and rather inexpensive. A typical use for these pads is under washers and dryers to keep them from shaking the baby awake during the spin cycle. I generally dry my babies on a line, but to each their own. Look up isolation pads made of a product called Sorbothane; this will provide good durability and excellent vibration absorption. Because home use is usually for washers and driers, your local appliance store may be a good place to check first.

Let's determine the combined weight of the necessary ingredients:

- Plywood; you need about 50 pounds for a $\frac{1}{2}$ inch 4 by 8 panel. Use two sheets to make a drum riser sandwich, for a total of 100 pounds.

- A sheet of mass-loaded vinyl or (cheaper) neoprene rubber. This will be the bologna in your plywood panel sandwich. Mass-loaded vinyl is a heavy, flexible sheet used for acoustic insulation in walls. This may weigh quite a bit, which adds to the dampening effect. Neoprene rubber is less expensive, but lighter, adding less sound-dampening mass to the riser, and more flexible, aiding the decoupling effect. If you need to move the riser on occasion, neoprene may be the best way to go.

- The drum kit (100 pounds, give or take)

- A drummer (let's say 200 pounds, including beer belly)

That gives you a total of about 400 pounds in most situations. If you use mass-loaded vinyl, be sure to figure its weight in as well. You only need a ballpark figure.

The trick is, you don't want to over-engineer this platform, because the vibration absorbing material needs to be somewhat loaded down to function optimally. You'll probably need pads rated at about 30 PSI (pounds per square inch), although the supplier will certainly be able to help you when you provide the weight details. Again, just be sure they don't say, "Well, you might as well make it good and strong, like, up to 1,000 pounds, just in case..." Don't over-engineer it, or you'll undo the purpose the project.

The setup is then quite easy. Figure 4.1 shows how it looks.

Figure 4.1 Building a drum riser out of vibration-absorbing materials can reduce the physical conduction of vibrations into your studio, leading to a clearer sound and happier neighbors.

Using an adhesive like liquid nails, adhere the vibration pads to one sheet of plywood. On the other side, affix the neoprene rubber. An aerosol adhesive is fine for this, just do it outside and away from your microphones!

Now put this in position in the studio, vibration pad side down, and lay the second sheet of plywood on top. It is best not to glue the layers together in case you want to move the riser about. Gluing also creates a less flexible connection, helping to transmit those naughty vibrations. Should the top sheet "wander" as the drummer plays, try adhering some strips of neoprene to the underside of the top piece of plywood. This should create enough additional friction to keep it in place.

Feel free to add a layer of carpeting on top for looks and to keep the kit from moving. This may be tacked or glued in place, as long as you do not nail through into the bottom piece of plywood!

Floor Treatment Tips

When you just wanna rock out, the mood can go down the tubes pretty quickly when the downstairs neighbor starts rapping on the ceiling with a broom handle. Unless you own the apartment, you won't be able to build a second floor to keep sound in, so you'll need to find a solution to reduce sound transmission without altering the building.

Although it comes as a surprise to many, simply carpeting a room doesn't help the acoustics much. Carpet on hard sub-flooring doesn't dampen physical vibration transmission effectively. It also absorbs high-frequency acoustic energy, leading to a muddy-sounding room, and tempting you to turn up the monitors even louder to hear clearly . . . a nasty cycle leading to deafness and eviction.

A better solution is to use a hard floor surface with an absorptive layer underneath. In this way, the firm top layer transmits energy into a broader area of the absorbing layer. This insulates your floor more effectively, and leaves you with a surface that reflects high frequencies for a brighter, clearer sound. I'll deal with the naughtiest reflections in the next section, so no worries there.

Just like the drum riser project, you want to create a sandwich that transmits energy very inefficiently. Figure 4.2 shows how you'll want to layer your floor.

The top layer is your parquet or laminate (cheaper) flooring. Beneath that are the underlay materials, separating the top flooring from the sub-flooring. This can be as simple as one layer of rubber/foam, but for maximum insulation, four layers can be used.

Under the laminate is the foam underlay. Next, you bring out the big guns with a layer of mass-loaded vinyl, which absorbs even lower frequencies. The bottom layer (just above the sub-flooring) is $5/32$-inch or $1/8$-inch cork. Cork is a superb material for floor sound

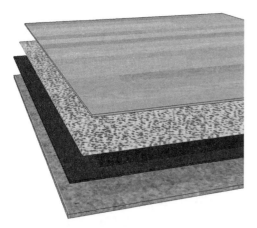

Figure 4.2   The top layer of your studio floor can be a wood or laminate layer, which helps keep the room sounding acoustically live. By using absorptive layers beneath the reflective top layer, you will reduce the transmission of sound out of the studio room. This isn't as effective as a completely de-coupled floor, but it is much less expensive!

insulation, as it is more resilient than synthetic materials—it won't compress over time and become ineffective.

For a rough cost estimate, let's figure on a room that's used for mixing and tracking vocals, keys, and acoustic guitar; a typical project studio room. Let's say that it's around 10×15 feet (about 150 square feet, or 14 square meters). For a 150 square foot room, the materials should cost as follows:

- *Laminate flooring:* Costs as little as $150. Many home construction suppliers have regular offers of $0.99/square foot, provided you're going for the budget options. This can naturally be more expensive for fancier materials.

- *First underlay (foam/rubber pad):* You should be able to find enough for 150 square feet for about $100.

- *Mass Loaded Vinyl (MLV for short):* This may very well be the most expensive element, it is, however, the most effective sound absorber. The "sound barrier" MLV from Audioseal weighs in at 1 lb./sq. foot, and will cost you about $250 for a 10x15 room.

- *Cork second underlay:* Usually around $150–$200, depending on thickness. Well worth the investment.

So, considering these prices, you could be set to rock n' roll with around $700 invested in keeping the neighbors off your back. Considering the costs involved in moving, this may turn out to be a worthwhile investment. With this method of creating a studio floor, you can take the materials with you if you move to another space.

Step 2: Controlling Resonances and Reflections

Having the doors, windows, and floors pepped up with the tips from the last section, and the amps and drums raised off the floor, you're on the way to a budget-recordist's paradise. Now, let's take care of the acoustic problems that are based on the physical shape of your rooms.

Just like a drum, guitar body, or any other space, a room resonates at particular frequencies, known as *modes*. Modes are not your friends, and neither are reflections. Reflections in small rooms with parallel surfaces lead to phase cancellation and comb filtering, as well as clearly audible "flutter" echoes. Have a look at Figures 4.3 and 4.4, which show the path of sound from your monitors, the problems that arise, and how you can handle these problems.

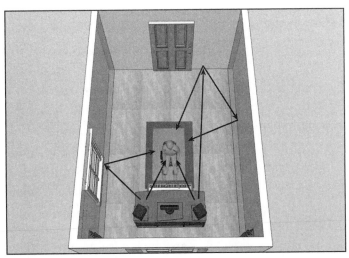

Figure 4.3   The path of sound waves arriving at your ears in an untreated control room. Note that multiple "hard" reflections of the sound arrive at the listener's head. This leads to phase-related problems such as poor stereo imaging, comb filtering, and a general inability to understand what the heck is going on.

So your duty is clear; stop resonances and lousy reflections, and you'll have a good-sounding room. Luckily, this can also be done in a modular way that allows you to non-destructively change the room to control these problems. Your tools are the following units:

■   *Sound absorbing panels:* These are inexpensive to build, broadband absorbers. This means that all frequencies are absorbed to a certain extent. In practice, lower frequencies are not absorbed as well as mid and high frequencies. Thicker panels are more effective, with the negative tradeoff of eating up more of your studio space.

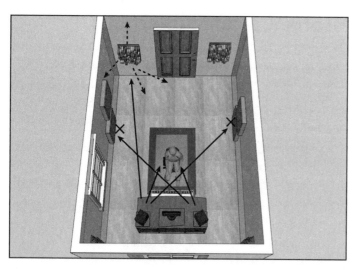

Figure 4.4 With a few simple sound absorbers and a bit of diffusion in the back of the room, you can significantly improve the listener's situation. Close reflections are absorbed by sound absorbing panels, and the slap-echoes from the back walls are broken up into a mild, natural reverberation.

- *Bass traps:* Low-frequency absorbing bodies that eat up "woofy" low-frequency energy that tends to build up in room corners. The simplest do-it-yourself design is cheap and easy, and will make a significant difference in the sound of your room.

- *Sound diffusers:* Since you don't want to absorb all the sound in your studio, turning it into a padded cell, diffusers "break up" reflected sound, reducing flutter echoes and phase problems, while simultaneously creating a greater sense of space in a small room. The downside is that they are rather expensive to buy, and quite difficult to build. You can create an effective room without any diffusers, so there's no reason to worry. They're also something that can be added at a later time without any difficulty, making a nice upgrade as your budget allows.

- *Resonant absorbers:* For particularly strong resonances in a room, you may have to build a sound absorber focused specifically at the offending frequency. By tuning the absorber to this frequency (by way of its dimensions), you invite in the unwanted sound energy. As the offending frequency resonates in the tuned cavity of the absorber, the sound energy is turned into physical motion and caught in the insulation. Because sound carries relatively little energy in the grand scheme of things, this physical motion is turned into a minute amount of heat. You've nabbed that nasty resonance.

Keep in mind: I'm not suggesting you use blankets, carpets, or egg cartons to acoustically treat the studio. These materials are to be used for sleeping, walking into the kitchen, and holding eggs, in that order. Don't believe what you see in jam-rooms and read on the web; these are not useful sound solutions.

The following sections discuss how to build these cool things.

Sound Absorbing Panels

These units are going to be your best friends. They absorb sound where you place them, turning a boomy, reflective room into a recording-happy environment in no time.

The easiest design involves Owens-Corning 703 rigid insulation. Basically, all you have to do is hang this stuff on the walls, and you're done. Just to make things look nice, and to hold the stuff together a bit, you can create a basic wooden frame to place the insulation in, and wrap it in some light cotton fabric to make it look neat. That's it! The 4-inch insulation absorbs more sound than the 2-inch stuff. If you have the extra space then go with 4-inch insulation, especially in corners where bass frequencies gather. See Figures 4.5 through 4.8.

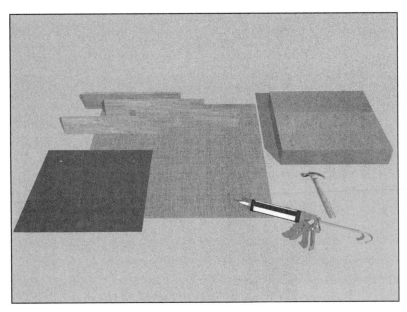

Figure 4.5 A simple wooden frame will be made to fit the insulation sheets. Light, thin wood is used to make wall mounting easier. The frame should be glued and tacked together, as screws could crack the wood.

If you can't get Owens-Corning 703, you can use rockwool as absorbing material (Roxul brand is typical). It is somewhat more difficult to work with since it is not rigid, although it is significantly less expensive. 703 does, however, have better overall absorption properties, so you're more effective sucking up sound within the space you sacrifice for the absorbers; the investment definitely pays back.

Figure 4.6   The wood frame is laid over the cloth used to wrap the absorber, ready to be tacked and glued together. Next, the insulation will be placed into the frame.

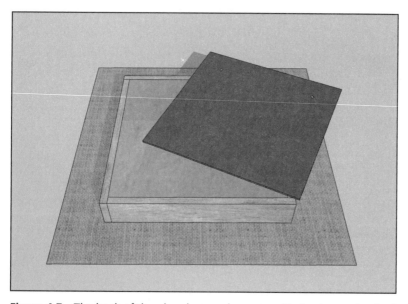

Figure 4.7   The back of the absorber can be covered in burlap, or backed with a thin wooden panel, as shown. If you plan on moving the panel often, a back panel lends it more stability, although the backside will then be reflective at higher frequencies!

Figure 4.8   Finished absorbers, ready to be used in your studio.

There's a reason that these simple broadband sound absorbers are a staple of the recording biz—they're cheap and they work. Make yourself a bunch of them, and you will not be sorry. This is one of the easiest "studio tricks" for improving your sound at the source.

Tip: The most problematic reflections in your space are the first reflections that occur as the sound leaves the sound source or the speakers, and is reflected quickly into the microphone or your ears, resulting in phase problems. By mounting a 2-foot by 4-foot tall absorber on a base, you can move it about to accommodate different instrument recording situations. Placing the absorber between the microphone and the closest reflective wall will result in a more solid-sounding track. For monitoring, do this: Place a small mirror right in front of each speaker, flat on with the front surface. Now, have a look into the mirror. The spot in the room that you see in the mirror is the first direct place sound hits after leaving the speaker. This is a great place to mount an absorber to reduce those close reflections and improve the accuracy of your monitoring.

The following projects become more time consuming, space consuming, nerve consuming, and expensive. They are not absolutely necessary, because most small studio acoustic problems are easily and most inexpensively solved with plenty of the absorbers just discussed. If you still feel that there are acoustic problems you want to pursue in your recording spaces, here are the next steps you should take.

Bass Traps

This may be even simpler than the sound absorbing panels, although they take up more precious space. By placing one of these babies in each room corner near your mix position, you'll pull some of the big peaks of bass energy out of your room's frequency response. Not only do these peaks make the level of bass instruments hard to hear clearly, you'll tend to overcompensate with EQ, and your mixes will sound lousy on other systems.

Here's what you'll need for this project. Head off to the home improvement store and buy two 32-gallon rigid rubber garbage cans, and a roll of mineral wool/rockwool insulation. You'll be cutting the insulation, so some long gloves and a facemask may be a good investment as well.

Assembly is simple. You measure the circumference of the garbage can, and cut two lengths of mineral wool to fit inside. Lay them inside the can, curled in there on top of one another so that the entire inside surface is covered with insulation. Now cut a piece to fit in the bottom of the can, large enough so that none of the bottom is showing. Make it a snug fit. Finally, cut another section of insulation to fit in a zigzag pattern into the middle space of the can, leaving some large air spaces, as shown in Figure 4.9.

Figure 4.9   A top-angle view of the garbage can bass trap, before placing the lid on the top. This unit creates a resonant/absorptive chamber to soak up rogue bass waves.

Now, place the top snugly on the can, and place your new bass traps in the corners of the room, preferably in the corners closest to monitoring position. As troublesome bass tones well up in those corners, the sound energy will be turned into mechanical energy inside the traps and absorbed by the insulation. Simple and inexpensive. As with the entire field of acoustics, there are far more complex problems behind the scenes than a garbage can full of rockwool can solve, but one point most acousticians and designers do agree on is that you can't have too much bass absorption. So when it's this easy and inexpensive, give it a shot, and then do some listening tests with and without the traps. You will probably find that the bass is "tighter" and more defined with the traps in place. If not, put the insulating material into a couple more wall-mounted absorbers and throw this book into the now empty trashcan. No, wait! Don't . . .

Sound Diffusers

Although they may look rather chaotic, diffusers are not just random slats or blocks of wood set up at odd angles to reflect sound all over the place. They are mathematically complex, and can be a royal pain to build. A D.I.Y. diffuser project can turn an otherwise pleasant Saturday afternoon into an aggravating collage of measuring, cutting, swearing, and gluing your hands to porous surfaces.

There's a fairly widespread belief that a bookshelf loaded with random books, trinkets, and forgotten vinyl albums functions as a diffuser. Well, it does and it doesn't. It is certainly better than a blank wall, but far less effective than a properly built professional unit. Diffusers have regularly occurring cavities of significant depth, designed to stagger the time it takes for various frequencies of the sound to reflect, making the reflection less coherent as a slap.

The problem with the bookshelf concept is that it's generally not consistent enough in the depth, reflectivity, and distribution of the cavities to make it effective. It is, as noted, better than nothing, acting more as a "scattering" device, and still helping prevent coherent flutter echoes at higher frequencies.

For you Saturday-afternoon masochists, here's a simple plan you can try. For the rest of you, consider ordering some finished units, should your budget allow. RPG Diffuser Systems offer some fine products, integrating bass traps into the units to absorb the sound they don't diffuse; a handy solution!

The following diffuser design is based on a BBC Research Department report called *The Design and Application of Modular, Acoustic Diffusing Elements*. The math involved in coming up with PRD and QRD diffusers designs would probably make you sob into your pillow. Just trust the numbers in Figure 4.10 and go with it.

So what does the chart in Figure 4.10 mean? It is the layout for the lengths of blocks to be attached to a stable backboard in order to create a three-dimensional diffuser, as shown in Figure 4.11.

The numbers shown in Figure 4.10 are ratios. The 0 means 0 inches. If the "1" equals 1 inch, your lengths are 1, 2, 3, and 4 inches. To make a diffuser effective down to around 800Hz, you should use 2-inch, 4-inch, 6-inch, and 8-inch blocks. This is going to be a

0	3	4	1	2	3	3	1	4	2	3	3
3	0	1	4	2	1	1	3	3	2	1	1
3	1	1	3	1	3	2	2	1	0	2	2
2	2	2	2	0	4	3	2	3	2	1	1
3	3	1	1	3	1	1	3	4	3	1	3
2	3	2	1	2	0	3	2	4	2	1	0
2	3	2	1	3	1	2	2	3	1	3	4
2	0	2	4	4	0	1	2	1	4	2	2
3	4	1	0	1	3	3	1	0	2	3	3
1	3	3	1	2	4	1	2	0	1	3	1
2	1	2	3	1	3	3	2	4	2	3	4
2	4	2	3	3	1	1	2	0	3	1	0

Figure 4.10   Is it a crossword puzzle for computers? Sudoku on mescaline? Nope. It's the layout for a sound diffuser.

Figure 4.11   What the diffuser looks like when all the pieces are cut to length and glued into place. This is one of the more difficult projects, but it can dramatically improve the acoustics of your studio.

heavy unit if you use 2 by 2 wood studs—somewhere in the neighborhood of 25 pounds. You may want to make several diffusers at once, since once you start all the cutting and gluing, you may as well have a pair for each side of the back wall opposite your monitors.

Here's what to do:

1.   Buy a 2-foot by 2-foot backing board, either plywood or hardboard. You'll glue the blocks to this in the pattern shown in Figure 4.10.

2.   Buy seven eight-foot lengths of 2-inch by 2-inch lumber. Try to eyeball them at the store to be sure that there are no bananas in there. You can have them cut in half for easier transport. You'll be doing plenty more cutting later. Don't forget wood glue, wood primer, and the paint/stain of your choice for the finished blocks.

3.   To get started, trace a 2-inch grid onto the backing board. Carry over the numbers as shown in Figure 4.10. Double-check your work, as there is some serious acoustics mathematics behind this, and if you mess it up, your mixes will all sound like remakes of "Tiptoe Through the Tulips."

4.   Having fun yet? If you are then you haven't cut all the boards to length. Get back to work! You'll need 38 pieces each of the lengths 1 and 2 (in this case, 2-inch and 4-inch). You'll also need 40 pieces of 6-inch boards and 15 pieces of 8-inch boards. When you go to cut them all, be sure to figure in the blade width, as you're always losing a bit of board.

5.   For a neater look, and better reflective properties, sand the ends and prime them. Do this on just one end, since the other will be glued down.

6.   You can paint the blocks, leave them natural, or stain them as you like. Staining them different hues according to length will make the finished diffuser look pretty snazzy. Don't paint the ends that will be glued. Keep the pieces sorted by length—you're entering the home stretch!

7.   Glue the pieces to the backing board as shown in Figure 4.11. Leave the "0" spaces empty, and check back as you go to be sure blocks are not pushing each other around. Wipe excess glue out of the "0" slots with a rag. Let this baby dry overnight before moving it around.

You're done! Mount the units on the back wall behind your mix position, or mount them on stands for mobility; they sound great placed in the tracking room as well.

I warned you that a diffuser is a pain to build! It's pretty easy to understand why finished units cost $250 a piece—it's not the material, but the labor involved.

You could theoretically make these out of some sort of stiff foam, but cutting the foam to size is probably fairly inaccurate and messy. The upside is that the foam would absorb some of the lower frequencies it doesn't reflect. The reduced weight is also a bonus, and would allow you to mount some diffusers on the ceiling, which is a great idea over your mix position or in the studio, especially over the drum kit position.

Resonant Absorbers

In general, you will probably be better off using broadband absorbers to take care of problematic bass frequencies, because small room acoustics are difficult to nail down without a degree in mathematics. You also really can't have too much bass absorption.

But if you happen to be in a fairly square room (that is, with even two of the same dimensions), you may need to attack the problematic room modes with more than simple absorbers. You need the big guns: resonators.

There is historical evidence that the Romans used clay urns as resonators to balance the acoustics of their amphitheaters, both to absorb bass frequencies and add "ringing" reverberation at desired frequencies. You have surely blown across the neck of a bottle to get it to resonate. You may have also noticed that the resonant tone becomes lower as you drink more beer out of the bottle. Well, you can use this same effect to take the wind out of detrimental room resonances by placing properly tuned resonators in your studio. The volume of the unit has a resonant frequency, and is individually designed to coincide with the problem frequency in the room. So take another gulp of beer for, uh, "tuning purposes."

The benefit to using a slat-type resonator in your studio is twofold; you target a more specific bass frequency problem, and the slats of the resonator reflect higher frequencies, maintaining some acoustic liveliness in the room.

First of all, you need to identify the problem frequencies. This can be done easily by using the *room mode calculator* spreadsheet included in Chapter 4 on the CD-ROM. Use the length, width, and height of your room to find the big bad modes, then you'll design some resonators to eat them up. The easiest resonators to build use a slat-type design, and the spreadsheet also shows you how to adjust the dimensions of the resonator in order to tune the resonator to your room.

So, let's say that you calculate that in your 10-foot by 15-foot (with 9-foot ceilings) room, you have the following modes:

- The 9-foot ceiling gives the following modes: 63Hz, 126Hz, 188Hz, and 251Hz
- The 10-foot width gives the following modes: 57Hz, 113Hz, 170Hz, and 226Hz

- The 15-foot length gives the following modes: 38Hz, 75Hz, 113Hz, and 151Hz

It so happens that the easier-to-design resonators are effective down to about 200Hz (due to size considerations), so let's focus on the 251Hz/226Hz problem area. You'll notice that those close dimensions—9-foot ceiling and 10-foot width—result in resonances fairly close in frequency. This leads to a nasty "bump" in that part of your room's tone. The good news is that you'll have about 15% around the designed effective frequency, so you're doing well even by building just one resonant absorber to take care of the 226Hz problem.

By including absorptive insulation inside the resonator, the effectiveness at the center frequency is reduced while broadening the bandwidth (Q) of the resonator. A wider Q means the unit will also absorb neighboring frequencies, which is exactly what you want here, since you have two problem modes—251Hz and 226Hz.

So, fire up the spreadsheet, and let's pop in a few numbers. You want to come out with 226Hz as the center frequency.

It's easy to plan on using 1-inch by 4-inch wooden slats, which shouldn't be too hard to find. The 1-inch slat depth also helps the stability of the construction, so the slats don't resonate and cause an audible problem. You'll be working with four variables:

- The *depth* and *width* of the wooden slats. If you've decided on 1-inch by 4-inch, these values can be entered in the spreadsheet right away. Be sure to measure your wood carefully, as actual measurements of the wood you purchase will vary. Enter the measured values for more accuracy—you will be able to adjust the other dimensions to compensate.

- The *width of the slots* between the wooden slats.

- The *depth of the resonator body.* If you anchor the frame directly to a wall, you need no backing material, but if you want to be able to move the resonator, you'll have to include a back panel. Quarter-inch plywood is suitable.

Figure 4.12 shows the end product.

With sound absorbing panels, a couple of diffusers, and perhaps a resonant absorber in place, your studio is now in top form for tracking and listening to your music. Keep in mind that even if you plan on producing music using samples, synths, and sound modules, the acoustics of your listening room need to be in order for mixing. Just because you don't plan on tracking pianos, guitars, drums, and so on doesn't mean that you can ignore room acoustics!

Figure 4.12 A slat-design Helmholz resonator, built using 1-inch by 4-inch wooden slats (shown with some slats not yet in place). This unit contains a layer of absorptive fiberglass insulation, which broadens its frequency range in trade for less absorption at the calculated frequency.

Considering Studio Ergonomics

Once the room is in order as far as acoustics go, you need to set up a comfortable work position, and wire up your gear.

First of all, be sure that you're comfortable in the control room listening position. Furniture, lighting, layout; these are matters of taste. Just remember that you'll be spending some long sessions in that location. Consider these points, using them as a checklist to prepare your workspace:

- *Lighting:* Indirect lighting is usually most pleasant, just be sure you have sufficient light, as this will help reduce eyestrain, headaches, and tripping over your own size 13s.

- *Furniture:* Get yourself a comfortable chair for the mix position. Also keep another chair handy for musicians/producers who want to move up close to the mix position. Beyond that, be sure you have ample couch space a little bit back from the mix position, so band members can hang out and make faces behind your back.

- *Access to the back of your gear:* Again, don't ram your mix position desk right up against a wall. Once you do that, Murphy's Law guarantees that you'll have to disconnect cable "A" and reconnect it to input "Y" or to input "Z." Especially if you don't have loads of gear, you'll probably need to have some preamps, compressors, and so on do multiple jobs. If you have things wired for easy access, you'll get the most out of your gear. Otherwise, you'll get lazy and that awesome gear won't benefit you as much as it can.

- *Monitor positioning:* Don't put the main monitor speakers right up against a wall, and especially not in a corner, which is a surefire way to excite room resonances. Place the monitors at a height where you won't have to slouch or strain your neck to get in the "sweet spot." Additional speakers for checking mixes can be placed in less critical positions if space is a factor, because you will refer to them less often.

- *Be aware of volume levels:* Your ears will betray you if you listen at too-high decibel levels—or too low! 85dB is an optimal average level for listening. A $50 (or less) Radio Shack dB meter is a great investment in your hearing's future. Many of us tend to crank the music up and up as we work. It's fun to check out the music cranked a little once in a while, but for your own good, don't listen for more than a few minutes at anything over 95dB. Additionally, low-level audio tends to sound weak in the bass and treble frequencies (hence the "loudness" setting on home stereos, which boosts bass and treble for low-level listening). If you do prefer monitoring at, say, 70dB, be sure that you occasionally crank it up to be sure you're not over-compensating by cranking the bass and treble.

- *Keep a clear line of escape:* Leave the mix position once in a while when working. Flop down on the couch, walk around, or go grab a soda while listening. Not only is it good to get the blood flowing again after $3\frac{1}{2}$ hours of vocal editing, but you will also hear things differently from different positions. Even if you prefer a messy, artistic, chaotic studio (I have an order to my chaos!), just be sure you can access the back of your gear, in case of problems. And the door to get out—take a break once in a while!

Connecting Your Gear

This process is actually pretty easy, as long as you follow a couple simple rules:

- *Keep cable runs short:* Long cables degrade the audio signal. This is another one of those aspects that piles up, track for track. This goes for mic cables and patch cables.

- *Use balanced (three conductor) cabling whenever possible:* Most modern gear uses ¼-inch TRS-balanced cabling for interconnection. Take advantage of the noise-rejecting properties of balanced gear, and don't substitute two-conductor cables (guitar cords) when you can use TRS cables.

- *Don't use a mixer on the way to tape (unless you have a top-notch unit):* The most common way novice engineers put tone-destroying circuitry in the way of their audio signal is through a multi-channel mixer (like those 12- or 16-input mixers from Mackie, Behringer, and the like). These kinds of mixers are okay for monitoring, but unless you have absolutely no other mic preamps available, don't use them before going to tape. The signal path is much better if you simply connect *microphone > preamp > tape*, rather than *microphone > channel preamp > EQ circuit > channel fader > bus fader > tape*. Less is more.

- *Take unused gear out of the signal chain:* This is an extension of the last point, that less is more. Even if you're using some high-quality standalone units, disconnect unused gear instead of using a bypass switch. If the unit has a "hard bypass," this is fine, it really takes all the circuitry out of the path. Be especially sure that you "hard bypass" digital units, as there's no need to send your signal through an additional A/D and D/A conversion before it goes to tape!

- *Beware of noise from power supplies and "wall warts":* Manufacturers of budget gear often skimp on the power supply to save money (transformers can cost big bucks) and the result is a studio full of buzzing, humming power supplies. Try to keep those clunky black "wall wart" external power supplies a few feet away from audio cables or from the transformer's magnetic field. Also, keep power cables to amplifiers and other gear with internal power supplies clear of audio cables for the same reason. Some rack gear is also best kept a rack space or two away from other gear; if you are hearing a buzz or hum that you can't pinpoint, disconnect each unit until it goes away, and then focus in on the perpetrator, isolating it from other gear as best you can!

- *Keep the gear accessible:* A patchbay is a great way to do this, although with many DAWs, the outputs of your gear can be easily sent to any input you need. This is more important if you use outboard EQs, compressors, and other effects. Use a patchbay to make their inputs and outputs accessible at the front side of your mix positions, and you'll get a lot more use out of those pricey toys.

INSIDE THE BOXES
Understanding Patchbays   One of the few exceptions to the rule that you should take the shortest cabling path for an audio signal is the *patchbay*. A patchbay serves the wonderful organizational purpose of bringing all those audio connections around from the back of your gear and up to a single interface where you may easily interconnect audio as you please. The convenience of not having to climb behind my gear is worth every penny.

The patchbay consists of two rows of cable jacks in vertical pairs. Cables can be connected to the front and the back. Generally, you use short cables to bring the inputs and outputs of the gear to the backside of the jack rows, and you can then make connections from the front with short *patch cables.*

The nice thing about a patchbay is that you can set it up so that connections on the back panel are *normalled* together. This simply means that the top row of the back is normally connected to the bottom row on the back, unless you insert a cable.

An example of a normalled connection is that you may always want the outputs of your favorite mic preamplifier (1&2) connected to the inputs of your recorder (1&2). So you connect the outputs of the preamp to the back top row; jacks 1 and 2. Connect the inputs of your recorder to the back bottom row, jacks 1 and 2. Your mic pre-outs are now normalled to your recorder's inputs.

Now, imagine that you want to put a compressor in between mic pre 1 and input 1 of your recorder to do some vocal recording. All you have to do is plug a patch cable into the front-top cable jack 1 to get the output of the mic preamp. Doing so *disconnects* the signal from the recorder input, and allows you to send it elsewhere. Now plug this cable into the compressor input. Take the compressor output and patch it into the front bottom row jack 1—it is now reconnected to your recorder input 1. If you think about the gymnastics this would require without a patchbay, you'll understand why a busy studio can't function without a patchbay.

A *normalled patchbay* is likely the most common wiring, although *half-normalled* and denormalled have their uses as well.

Denormalled is easier to define. There is simply no normal connection from top to bottom; only from front to back. I have a denormalled patchbay for effects devices and compressors, so that I can keep the inputs and outputs appearing neatly over each other without causing a feedback loop (output connected to input—yeaoOOOOWW!). This is useful just for organizing things.

Half-normalled is just like normalled, except that plugging in a cable does not disconnect the top back row from the bottom back row. This can be useful when you need a copy of a signal for something like a compressor side chain to create a de-esser.

Abnormalled is what you will become if you actually try to memorize all this patchbay crap in one sitting.

(Text is excerpted from *Pro Techniques for Home Recording,* a previous work of the author.)

Studio Monitors: Price and Practicality

Among the toys that gear companies would like you to buy, studio monitors are a favorite. "Brand X, model Y monitors are great monitors" is a very subjective statement. How are they great? Do they sound accurate? Or do they just sound flattering? Is a wide frequency response best? Or would you rather hear something that sounds realistic—like consumer's stereos?

The truth is that all these things are necessary, so you need more than one set of monitors!

You should listen to your mix on every set of speakers you can get your hands on. Only then will you know if your mix sounds good on many speakers.

So, how much do you need to spend on monitors? Is there no hope for a good mix without speakers costing $2,000 a piece?

Again, there are so many companies offering monitors that competition is working for you, and there are plenty of quality choices that won't break the bank. Some relatively inexpensive models from Mackie, JBL, and KRK are conveniently self-powered, saving the added cost of a power amp for the speakers. There are generally many sizes offered, depending on the speaker sizes. An 8-inch woofer in your main monitors is a good idea, providing deeper bass response than more compact models.

The "problem" with great, expensive monitors in inexperienced hands is *precisely* that they sound great. Full-range, smooth-sounding monitors can reproduce everything you send their way, whereas a cheap boom box will sound good only if you've got your mix right. In order to get your mix to that level, you will want to hear the detail that only good monitors can reveal, but you should constantly check on headphones, cheap radios, and even TV speakers to be sure you're on the right track.

A typical problem with mixes done only on the "great" monitors is cranked bass and high treble. Because you can really *feel* that luxurious bass, and hear all those clear frequencies up to 20kHz, the tendency is to enhance this effect, which doesn't translate well to other types of speakers. Checking on some lousy speakers is a good way to ensure a great mix—if it sounds good on a blown-up boom box, a TV, a pair of consumer stereo speakers, and your studio monitors, then you're on the right track.

Eventually, you will be able to work on your best monitors with less need to constantly compare—you will get used to how things need to sound in order to translate well to other systems. To help the learning process along, spend a lot of time listening to your favorite music on your studio monitors. In fact, if you're mixing a song with similar qualities to a professionally done song you have handy, compare your mix to theirs. I'll get deeper into this issue in Chapter 11, "Mixing: Balancing Art and Craft;" just be sure to set up a few extra speakers now as you are putting your studio together.

Final Thoughts

Setting up your studio is not a project to be underestimated. Much of the quality of recorded tracks depends upon the acoustics of the recording room, as do the acoustics of the control room influence processing and mixing decisions. Time invested in this stage will pay you back during every project you work on in your studio.

So, now that you've taken the first steps in creating a pleasant-sounding and pleasant-to-work-in recording space, and wired it correctly, you need to learn how to swing a microphone into the right place to capture the tone you want. This is both an art and a science. The next chapter will get you up to speed on the most common methods engineers use, and reveal a few tricks along the way.

5 Microphone Technique

With a good-sounding source set up in your good-sounding room, it's now time for you to bring some engineering skills to the party. In some cases, solid microphone technique will be your most important contribution to the project. No amount of creative mixing and expensive outboard gear can overcome poorly recorded tracks; this is something you have to get right from the start.

Allow yourself some time at this stage, and be sure your client knows that this is needed. In professional studio circles, it is usually accepted that for a good drum sound about an hour is necessary for setting up mics, adjusting levels, listening back, readjusting, and listening again until the producer is happy. One- or two-mic instruments like acoustic guitar, sax, percussion, vocals, and so on should be ready to roll pretty quickly, and having your studio wired for action will help you get there.

Part I: Setting Up

First and foremost you need to get a clean signal at the right level into your recording device. I'll use the term "to tape" when referring to tape, hard disk, recordable media, and so on, any sound signal being recorded will be going to tape for the sake of easy reading. Otherwise you'll be reading, "Before sending the signal to tape/hard disk/flash drive be sure that you check the blah blah blah..." "To tape" is simply an old expression from analog days.

In the case of wiring your studio, there *is* an objective best way to do it. This is the way that gets the signal to tape with the least amount of wiring along the way. Every time a signal passes through more resistors, transformers, capacitors, toaster ovens, and miles of cable, it is degraded. It doesn't matter if you had to sell a kidney to afford that fancy, custom outboard equalizer; if it is not being used to process the signal, take it out of the chain! The truth is that no piece of gear is going to make the signal better. In the best case, an outboard audio unit will control dynamics or correct an EQ problem in a *minimally lousy way,* but it won't actually improve the quality of the signal coming out of the microphone as far as accuracy goes.

If you want a particular effect, and you are sure that you won't want to remove it later, then by all means print the effect to tape. I often print compression/limiting on drum room mics to tape if I find a sweet spot. Same goes for unique EQ effects on filtered guitars, vocals, keys, and so on. If the effect is part of the performance, don't lose it, and be sure to write down your settings in case you choose to retake something at a later time.

So to sum up proper wiring for tracking, the signal should flow like this: Source > Microphone > *shortest possible cable* > (Perhaps an outboard EQ or compressor)> *shortest possible cable* > A/D converter > To tape/disk.

INSIDE THE BOXES

Cables Actually Function as Filters   *Cable capacitance* can be measured in any cable, and the longer the cable, the more this effect will filter high frequencies (read: clarity and definition) out of your sound. Don't be fooled by overpriced platinum-infused boutique cables, just stick with the shortest cable you can. If you are tracking a shaker overdub in the control room, disconnect that 75-foot cable that leads into the mic cable box in your tracking room, and just plug a six-foot mic cable right into the preamp; it does make a difference!

 In Chapter 5 on the CD-ROM, listen to the first example provided. The example is a shaker recorded with two microphones—one through a hundred feet of cable, and one recorded through just six feet of cable. For ease of comparison, the shaker was recorded simultaneously through the same microphone type, preamp type, and stereo converter. The only difference is that one side ran through loads and loads of cabling, whereas the other side through just a few feet. I've panned one to the left and the other to the right, at the same level. You will notice that the shaker doesn't seem to appear in the middle; the loss of high frequencies on the long-cable side makes it seem like the shaker is panned toward the "clear" side.

Keep this in mind whenever you connect gear. Overly long cable runs are a common home studio mistake; consistently using the shortest possible cable will increase the clarity of your recordings.

This *signal chain* sometimes consists of standalone units, but many recordists use an all-in-one box, like a Pro Tools MBox or 00x series, RME Fireface, M-Audio Fast Track, Presonus Firebox, or other brand. These are the type of units connected via FireWire or USB directly to a computer, and vary in the quality of the internal microphone preamps and A/D converters. RME happens to make some very good converters, on par with units costing significantly more, and are a good recommendation for a fledgling studio.

If you are looking to invest in the quality of your recordings, consider improving the components in the signal chain that lead to tape. Any professional studio relies on its preamplifiers and A/D converters for a sound that captures a rich, detailed sound with a minimum of noise and other audio artifacts.

Let's say that you start off with a Pro Tools 002 interface, for example (I'm trying to remain brand-neutral here, but someone needs to step up to the whipping post...). You would simply connect the 002 interface to your computer via a USB cable, start the software, and be ready to make some tracks. The only problem is, your audio is at the mercy of the components built into that unit.

Without getting into the workings of electronics, a single high-quality preamplifier requires components, which usually cost as much as that entire interface costs. To make multi-track recording affordable to a larger consumer base, the companies that build these units are forced to cut corners. The symptoms of these engineering shortcuts are usually most apparent as the tracks start piling up.

 In the next example on the CD-ROM, you can listen to two recordings, one made with a single, inexpensive all-in-one recording interface, and one with a pro-quality chain including a high-quality preamplifier and A/D converter.

By comparing the two examples, you should clearly hear the advantages of using a pro-level chain of audio devices to get your signal to tape. With preamplifiers costing $500 or more per unit and proper A/D conversion at $1,500 or so for eight channels, this may be more of a long-term goal. Regardless, it is important to know the ingredients needed to get that "big studio" sound. You may want to rent or borrow this kind of gear for an important session. I find it worthwhile to cut drum tracks in a studio fitted with this type of gear. Enough preamps and converters to track a whole band would be expensive to rent, and the studio also offers better monitoring and a great microphone cabinet.

A few tips for choosing an interface:

- The market is fairly flooded with preamps, converters, and the all-in-one units, so choosing an interface can be difficult. Consider how important portability is for you, and how many tracks you need to record simultaneously (how many inputs you need).

- If portability is a key feature, it is easy to carry a laptop and a small interface with you. The "studio to go" is certainly possible now, although when recording acoustic instruments, the interface's sound quality may still be a compromise. Apogee offers the Duet and Solo interfaces, which include their top-quality converters, making these interfaces a great option for portable recording when audio quality is crucial.

- If you plan on recording a whole band, including drums, you will need more than the typical two (maybe four) inputs on a small all-in-one box. In this case, I recommend an RME Fireface 800, Apogee Ensemble, or similar high-quality multi-channel converter interface for non-Pro Tools systems. Pro Tools users will have to use a DigiDesign interface, and may want to consider letting the folks at Black Lion Audio (look them up on the web) overhaul their interface for better audio quality. You could also use an external converter and feed the A/D output into the digital inputs of your DigiDesign interface. In my experience, these are common solutions that budget- and quality-conscious studios have opted for in recent years.

■ The good news is that you can improve the links in your chain one by one. The most important place to start is with the microphone preamplifier. The difference in audio quality between a stock preamplifier on an inexpensive mixer or all-in-one interface and a Class A solid state or tube preamp is impressive.

Part II: Understanding Stereo Microphone Techniques

Sticking a single microphone in front of a sound source and pressing record may capture the performance for posterity, but to your two ears, this type of recording often lacks depth and realism. Since the 1960s, engineers have been experimenting with methods using two microphones in order to record sound in a way that can fool your ears into thinking you're hearing a live performance with a sense of depth to it. There are numerous ways to create stereo recordings with two microphones, and a variety of techniques have been developed for different situations.

Stereo recording is one of those topics whereby engineer hotshots like to throw around terms like Blumlein Array, ORTF Pair, Mid-Side Matrix, and Ph.D. Frink's Altitudinal Variatory Segmentation. Well, the last one is made up, but you get the drift—there's a lot of crap to be waded through.

Basically, the concept is simple: you take two microphones, put them near the source, and put the recording of one in one speaker, and the other in the other speaker.

XY Configuration

The easiest way to get started recording in stereo is to set up two mics with their capsules close together, in what is called an *XY configuration,* as shown in Figure 5.1. This is easiest with small diaphragm condensers, due to the capsule size, and works great in a small room, where you might not want to pick up much ambient information, and just want a tight image of the source. It's great for drum overheads and stereo acoustic guitars in a small room situation. The XY configuration is also referred to as a *coincident pair.*

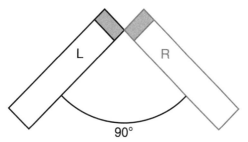

Figure 5.1   XY microphone configuration.

If you use two figure-eight pattern mics for this, it is known as the *Blumlein Array*. There you go—another bit of jargon swept aside. The Blumlein setup is best when placed back farther from the source, as the stereo image does not appear to be as wide. XY setups are highly mono compatible, so if you expect your recording to be played on a standard TV, this is a good way to go.

Tip: Although you can use two separate stands to position the microphones, it is well worth the $20 investment in a stereo mic clip to stabilize the position of the mics. Should motion of the performers vibrate one of the mic stands, phasing sounds could result.

ORTF Configuration

A slight variation on the XY technique is the *ORTF* technique, which stands for Office de Radiodiffusion Télévision Française, where the technique was developed in the 1960s. This setup calls for two cardioid pattern microphones to be set up with the capsules 17 cm apart, and at an angle of 110 degrees. This technique creates a wider stereo image, and is still relatively mono-compatible. The *NOS technique* calls for a separation of 30 cm and 90 degrees. Just another option. You can experiment with the angle and distance, although the ORTF and NOS spacings and angles have been calculated by audio professionals wearing impressive hats.

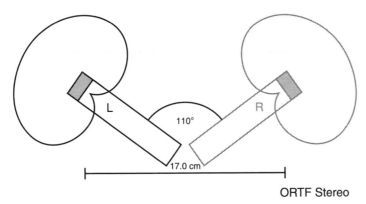

ORTF Stereo

Figure 5.2   ORTF stereo mic setup. This will provide you with a wider stereo image at the expense of mono compatibility.

Spaced Pair Configuration

The *spaced pair* technique is just that—a pair of microphones set apart from one another, as illustrated in Figure 5.3. For orchestral recording, a spaced pair of omni mics provides impressive stereo imaging and room ambience at the expense of mono compatibility. This is also a common way of setting up room mics for capturing drum ambience. Be sure to make a test recording, and be sure that the reverberant sounds are not imbalanced. I have run into this problem before when recording in small rooms, and it can make the drums sound skewed to one side.

The other mic is over there!

Figure 5.3   The spaced pair stereo recording technique. If you're reading this at home, the second microphone is in your kitchen.

Baffled Pair Configuration

A very clever way of recording stereo involves *sacrificing a human head*. I thought that might get your attention in case you were nodding off. It is actually much less evil than it sounds, since a foam head or just some kind of baffle is used. Actually, a thick pillow will do in a pinch, as shown in Figure 5.4.

When setting up a baffled pair, the mics should be set to omni mode—you're basically simulating two ears and a head in between. The stereo image is particularly sharp, and ambience is more defined than with the XY, ORTF, and NOS methods. It just looks a little sillier. I think it's a great option, because it is also easy to set up using large

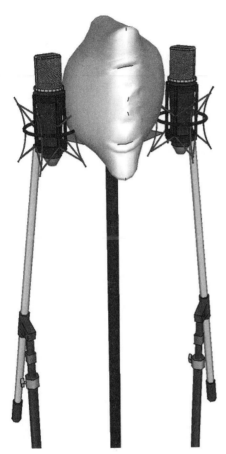

Figure 5.4 This is known as the baffled pair technique, referring to acoustic baffling between two microphones. This simulates a listener with a head full of thick, sponge-like material; an accurate representation of an FM radio listener.

diaphragm condensers. Be sure to use a stable mic stand if you rig up a heavy pillow and two expensive mics and place the whole shebang in the same room with musicians. I would suggest busting out the gaffer tape, taping both stands together, and taping the stand's feet to the floor.

M/S Array Configuration

My personal favorite is the voodoo-like *M/S array,* which mysteriously creates a stereo image from two mono mics right on top of one another. This one requires one cardioid and one figure-eight mic, placed with the cardioid capsule facing the sound source, and the figure-eight mic looking to the left and right, its capsule right over the cardioid mic. M/S refers to mid-side, meaning that the cardioid mic is picking up *middle* information, and the figure-eight mic picks up the *sides.*

Optimally, you should use the same microphone make and model, just with a different pattern selected. Lacking such a selection of mics, you can use what you have available and it will still work well.

The voodoo is the way you have to set up a *decoding matrix* (another smarty-pants phrase) to create the stereo image. It works as illustrated in Figure 5.5.

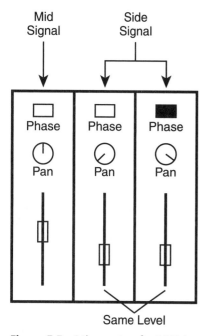

Figure 5.5   Mixer setup for M/S decoding matrix. This is easily set up using a DAW mixer, and can be saved in most programs as a template, saving setup time and encouraging you to use this great-sounding stereo technique.

The cardioid mic is panned to the center. The figure-eight mic signal is copied to two channels. These are then panned left and right, and the right side is polarity reversed. The left and right channels should be kept at the same gain as they are blended into the mix.

As the L/R channels are mixed in, a clear and mono-compatible stereo image appears. This method sounds particularly three-dimensional, and can be used to great effect in front of any sound source, from drums to vocalists. Because many recordings are being made with a single cardioid pattern mic in the modern studio anyway, why not add that figure-eight mic and be ready to pull off some 3D mix magic?

Part III: You May Now Touch the Microphones

You've made it. I am now going to turn you loose on actual audio sources. No matter what happens now, remember just one basic rule—always use your ears. Walk around near the performer and listen. Put a microphone where it sounds good, and make a test recording. This is a great way to start, especially if you're dealing with a new instrument, although you'll now learn some typical settings to fall back on (you may not want to stick your ear inside a bass drum to find out how it sounds...).

Each instrument mic'ing section that follows is broken into "Recommended Mics," "Typical Setups," "Try This" (for some unique sounds that might not fit your project...), and "Potential Problems." These tips go beyond the basics and include some tried-and-true studio tricks, but if time permits, be sure to experiment! Ultimately, any microphone works on any source; it just comes down subjective to sound quality. If a $10 Radio Shack CB microphone sounds good, it is good.

Mic'ing Vocals
Recommended Mics

Pretty much anything goes when mic'ing vocals. First choice is any large diaphragm condenser, or a simple dynamic mic (Shure SM-57 or 58). Good news is that even with a simple two-channel interface, you can record two mics at the same time. Try putting up a large diaphragm condenser with a Shure 57 next to it. You can A/B the two, blend them, or process one for effect. If one mic sounds particularly good, you can just go with that one and save tracks if needed. If neither sounds so great yet, swap out the lesser of the two and try another mic. This is an effective way to find the best mic for a singer.

Generally, the only mic type that won't work so great is a small diaphragm condenser. They tend to lack the warmth that large diaphragm condensers have, although don't rule them out completely. They can be great for background vocals, and I have had luck on lead vocals as well, having a singer just "give it a quick try." It turned into the final take.

Typical Setups

First of all, the singer should be comfortable. Most singers will prefer to stand, as it makes breathing easier, but if they choose to sit, adjust your mic stand accordingly. Be sure you have a pop filter and, if at all possible, a shock mount for the microphone. Many singers tap their feet or move about when singing, and this can carry up the mic stand into the microphone as thumps and rumbling.

If the singer needs to read from a lyric sheet, set up a music stand at eye level so they don't lower their chin and strangle the words. Set the lighting to a comfortable level. Some singers require strobe lights and lava lamps, others just want the light to be bright enough to follow the lyrics and not be shining right in their eyes. While setting up your studio, a trip to the nearest flea market should provide you with a few different lamps for a few dollars.

Set the microphone(s) 4–8 inches in front of the singer, with a pop filter an inch or two from the mic's own windscreen. For singers who heavily pop their Ps, you can try taping a pencil vertically down the middle of the mic's windscreen, as illustrated in Figure 5.6. This will split those plosives away from the mic diaphragm while minimally affecting the tone. This is even more important when using ribbon mics so as to avoid damaging the ribbon!

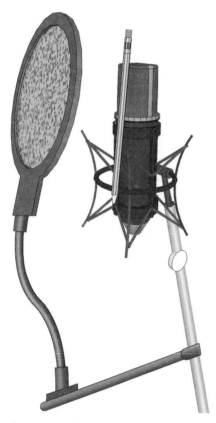

Figure 5.6 Setup for vocal recording with filter to control popped Ps, and the "pencil trick" to reduce sibilance.

Adjust the microphone preamplifier gain to set your recording level. Try to average around −18dBFS on your DAW's meters. The actual level will vary greatly during most songs, but this setting places the recording level safely below the danger of overloads and above the noise floor, especially when recording at 24 bits. An experienced singer will know how to play the mic, moving closer for soft passages, and backing away slightly when belting out a chorus, thereby controlling their own level to a degree. When working with inexperienced singers, you may want to put a compressor or limiter in the audio chain when tracking, set to catch surprise peaks, and prevent digital distortion. You'll find more on compression in Chapter 7, "Signal Processing Toys: Compression."

For less experienced singers, tell them to start off by staying a hand's width away from the microphone, and to otherwise just relax and let 'er rip. No need to push the technical side of things onto someone who may already be overwhelmed with the studio experience.

Try This

Aside from the trick of tracking with two mics at once, you can also expand your potential sounds when mixing by placing a microphone a few feet away from the singer to capture a broader image of the sound. This is a fairly common studio technique, and the mic is also often placed several feet up above the singer, catching some room tone along with the performance.

To help inspire a performance, you may want to experiment with effects. I find that amp simulators, filters, and/or heavy compression can send the vocal into sonic orbit, inspiring the singer to perform differently. The best way to do this is to use effects from plug-ins that won't be permanently recorded to tape. You can set them up on an effects send, or just have them set up on the monitoring channel. Be sure to save the settings, because the tone will surely become part of the performance, although amp distortion and heavy compression typically needs to be backed off a bit when it comes time to mix. For this reason, it's important not to record the effects to tape, but rather have them on a monitor-only channel, and, again, save those settings!

Some singers can't get into the right mood with headphones on, and a patient engineer, resisting the natural urge to strangle the singer, once developed a trick that allows you to use monitors to record the vocal, while still keeping the recorded music from bleeding into the mic! By using sound wave cancellation to your advantage, you can set this up. The trick is to set up two monitors and the microphone in an *equilateral triangle* (remember that from geometry class?); which means that the distance between the speakers, and from each speaker to the mic is the same. Measure from the center of the speaker cones, and the diaphragm of the mic, as shown in Figure 5.7.

The setup is fairly easy; there are just a few things to observe:

- *You must use a mono mix of the backing track, same in each speaker.* Since the same mix is used, by flipping the polarity of the mix in just one speaker causes the two speaker's outputs to cancel at the microphone. The easiest way to do this is to make a mono sub-mix of the music, and copy it to two tracks. Mute all the other tracks. Pan one mono mix left and the other right. Now, flip the polarity of the right mono track using a plug-in or processor. Set up a track for the vocal, but keep it muted.

- *Be sure that your distances are measured accurately.* Try three feet between the speakers, and from each speaker to the mic. That should be a comfortable distance.

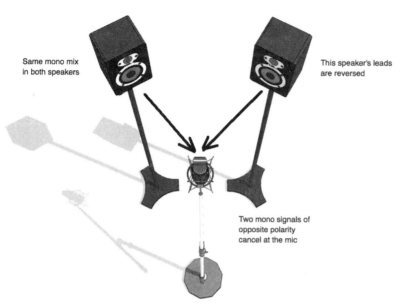

Same mono mix
in both speakers

This speaker's leads
are reversed

Two mono signals of
opposite polarity
cancel at the mic

Figure 5.7 The "polarity-reversed triangle" singer setup. The singer hears the mix, but since the polarity of one speaker is reversed, the backing track is cancelled out at the microphone, leaving just the recorded vocal. More acoustics magic!

- *Placing some sound absorbers behind the singer will help reduce reflections from the speakers, which won't cancel out at the mic.* Don't sweat a little bleed; the idea is to get the singer into a great performance, which is much more important than a pristine recording.

- *Don't add the vocal into the monitors, or you will cancel it out as well!* You may also be in for a blast of feedback if your triangle is not perfect . . . If the singer wants to hear some of the vocal, try sending that to a headphone mix with whatever reverb they like. They can then play around with the headphones on one ear, and so on, to get comfortable. By using a "pre-fader send," you can send the vocal to an auxiliary output even if the track is muted.

Potential Problems

The most common problems when recording vocals are pops (from plosives like "B" and "P") and sibilance (sizzling sounds from "C," "F," and "S"). Pops are usually reigned in with a pop filter and the pencil trick, but editing may be needed after the fact. A low-cut filter is built into many mics, although be cautious of using this as a default setting; some male voices may lose depth if the cutoff point happens to be rather high. I prefer to find the few problem points in the vocal track and manually edit them if needed.

Tip: After you have your final take, set up an EQ plug-in on the vocal track. Loop an example section where the pop shows up. Find the problem frequency (generally a blast around 50–80Hz and below) and cut it until the pop becomes less noticeable. 6–10dB of cut may be needed. Now, automate the EQ to cut only where the pop occurs. This will keep the processing from affecting the tone where there was no problem. Next, listen in context. If the problem is well solved, you may want to process this edit back to the audio track permanently to save CPU power or plug-in instances. (Note: If this confused you, you may want to jump ahead to Chapter 6, "Signal Processing Toys: EQ," and read the section there called "equalization techniques" to learn some terminology.)

Sibilance is just as distracting as pops, and can actually hurt people's ears when playing back at a rockin' volume. *De-essing* is the name for getting rid of sibilance, and you'll look at that more closely in Chapter 6. In the meantime, if you have a particularly sibilant singer on your hands, try having this person sing slightly over the top of the microphone (that is, you aim the mic more towards the throat), or raising the mic slightly so that the bursts of sibilant Ss go slightly past the microphone's diaphragm. This may prove to be less noticeable than the action of a de-esser plug-in.

Mic'ing Piano
Recommended Mics
You'll probably need two mics to accompany the physical size of the piano and to provide some stereo imaging. For rock/pop recordings, large diaphragm condensers are a good starting point (AKG 414, AT4050, or Shure KSM-32). For classical music, small diaphragm condensers can provide more detail and room tone clarity (Shure SM-81, Neumann KM-184, and Brüel & Kjaer 4000 series). Ribbon mics' flat frequency response and excellent sensitivity to transients makes them a great choice for piano. For good old rock n' roll boogie piano, try a couple of Shure 57s or Sennheiser 421s; that may be all you need!

Typical Setups
For piano, the "use your ears" rule absolutely comes first. In a classical concert hall, this would be the best approach. Place a stereo pair of small-diaphragm condenser mics in the position where you hear a pleasing balance of the piano's acoustic character and the ambience of the hall (refer to "Part II: Understanding Stereo Microphone Techniques"). Alternately, you can place a pair just above the pianist's head, essentially picking up the performer's perspective. This may not be an option, however, when performing for an audience. Due to the complex acoustics of a grand piano, including the options for positioning the lid, an entire doctoral thesis could be prepared (and probably has been) on the subject. Let's focus more on recording a piano to fit into a larger production, instead of as a classical instrument, which is the scenario you are more likely to encounter.

The "go to" microphone in this case is the AKG C414, but any similar large-diaphragm mic with a fairly flat response will work. Avoid mics that have a high frequency boost. Due to the complex harmonics of a piano, this can affect the overall balance of the sound. If the situation allows you to leave the lid of the piano open, and the drummer isn't bashing away nearby, an omni pattern mic will pick up a more balanced tone. I've had decent results placing a pair of mics inside an upright piano with the lid closed, but to keep the sound from becoming boxy, the mics had to be placed into cardioid mode.

When close-mic'ing a piano, ask the player if he or she is going to be playing in the extreme upper and lower register of the piano. It is likely that she won't be hammering away in the lower octave, so if you use a spaced pair of mics, you may be able to bring the two mics closer together for better coverage of all the strings. Using a spaced pair (again, see "Part II: Understanding Stereo Microphone Techniques") is another possibility, but be sure to check for phase problems between the two mics.

If you are recording a grand piano, where the strings lay horizontally, you will also want to consider how far along the strings from the hammers you place the mics. Near the hammers, you will hear more attack, and the tone will be brighter. At the far end, the attack will be minimized and there will be considerably more sustain to the sound. Recording both of these mic positions allows for more flexibility when mixing. See Figures 5.8 and 5.9.

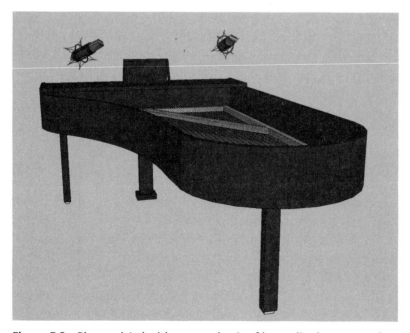

Figure 5.8 Piano mic'ed with a spaced pair of large diaphragm condenser microphones. This provides a wide stereo image.

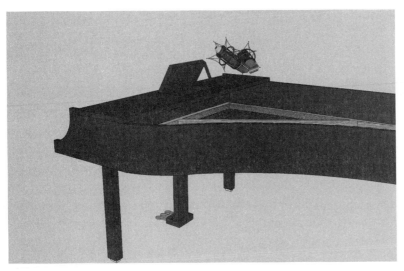

Figure 5.9   A piano mic'ed with a near-coincident pair of microphones. Some care is required to avoid missing notes that fall outside of the mic pickup pattern, and to be sure that there are no phase problems.

Try This

For pop and rock, compression is often just what the doctor ordered. By placing the mics fairly close to the hammers, you get the brightness and attack needed for a mix that also includes drums, bass, guitars, strings, and so on. Compression helps you reel in the attack and increase the level of the sustain portion of the sound. Another famous Beatles moment is the last piano chord in "A Day in the Life;" performed *fortissimo* and compressed so heavily that the release of the compressor allows the chord to ring for over 40 seconds. Serious sustain.

Potential Problems

Some cardioid microphones negatively color off-axis sound, or have a pattern that is particularly tight, making notes outside of the mic's pattern sound weak or smeared. Know your microphones, and chose accordingly! It is a serious bonus to have microphones in which the pickup pattern can be controlled from the power supply. This allows you to find a good position, and then try varying the pattern to see how it affects tone.

Bleed from other instruments is a consideration when placing two large diaphragm condenser mics in the room! If you're using an upright piano, it may be possible to roll it into an isolation room. If the band is going to rock out together and keep the takes they cut together, you may not have to worry about bleed too much. If the player wants to replace their performance later, the best modern solution to bleed problems when recording piano is to use one of the many excellent synth or sampled sound banks available. Problem solved.

Mic'ing Acoustic Guitar
Recommended Mics

Small diaphragm condensers: The Shure SM-81 is one of the best-known mics for acoustic guitar; it has been on so many recordings that it has become "the sound." For those on a budget, the Rode NT5 or MXL 603 are a good choice. You can grab a pair for less than $400, and they can also serve as drum overheads and room mics. In the mid-market range, the Oktava MC012 and Peluso CEMC6 give even upscale SDCs from Neumann, AKG, Schoeps, and other "big names" a run for their money. In the project studio environment, the 2%–3% quality increase that the priciest mics offer may become lost in poor room acoustics, less-than-optimal A/D converters, and the general imperfections in the home studio. Again, concentrate on performance on a decent instrument, and don't sweat the gear too much.

Ribbon microphones are an excellent choice for acoustic guitar. However, in my experience, a particular ribbon mic and acoustic guitar either match well, or don't get along at all. Guitars and ribbon mics all have their own "voice," and although finding a match can be a bit of work, it can be well worth the extra effort. Try any make or model of ribbon mic, and be critical of the tone. If the sound is very bass heavy due to the proximity effect, back the mic off a foot or so, and see if the sound opens up. Don't be afraid to swap an expensive mic (such as Coles, AEA, or Royer) for a cheaper one (Cascade, Nady, Golden Age, and so on)—you may be surprised!

Any *large diaphragm condenser* or *ribbon mic* is also worth auditioning. If you have a house acoustic guitar, it is good to get to know which mics like it best, and make these your standard. *Dynamic mics* would be the last choice in this case, as they have a relatively slow transient response. But trying never hurts!

Typical Setups

When recording acoustic guitar, consider the sonic space you have available in the mix. In many cases, a mono recording is sufficient—just walk around in front of the player, listening for a good spot, and place a mic there. Don't be afraid to back the mic away a little bit unless you want a particularly in-your-face presence to the recording. A little distance can really help create a sense of depth in the recording.

A stereo recording of acoustic guitar can be made in a few ways. Please refer to Chapter 5 on the CD-ROM to see and hear how different mic setups sound in more detail.

Try This

Connect a mic and put a track in monitor mode. Put some headphones on the guitarist, and have some ready for yourself. Roll the track, and have the guitarist play along. Now, move the mic around in front of the guitarist until you find a spot where the guitar tone comes into focus on the track. This is one of the best-known and most effective "big studio tricks" of all time. This also works great for mic'ing guitar amps, setting ambient drum mics, and dealing with instruments you may not have recorded before. By adjusting mic distance, height, and angle, you'll be EQing the recording to fit in the most natural way.

Typical advice is "always put on new strings before you record acoustic guitar." Again, use your ears; if the tone of the old strings is fine, don't bother changing the strings. A friend of mine has a Gibson J-45 that sounds awful with new strings; it's just too bright and clanky. As the strings get worn in, the guitar sounds best, so that's the way to record it!

If placing microphones in the typical close mic'ed positions isn't doing it for you, try going for the player's perspective again. Try a stereo pair of small diaphragm condenser microphones placed just above (or beside) the guitarist's head to pick up what they are hearing. If this still isn't satisfactory, you may need to consider a different guitar (or a different guitarist).

Potential Problems

The most common problem when recording acoustic guitar is picking up some sort of resonant boomy frequency in the low midrange that obscures the detail of the strings. Although this can be cured with EQ, try to find a mic position that minimizes this effect before you start twisting knobs. If you are close-mic'ing the guitar with a large diaphragm condenser, try moving away from the sound hole of the guitar. Also consider using a baffle near the guitar to absorb room resonances, which may be where the "boominess" is coming from.

Mic'ing Electric Guitar
Recommended Mics

Here's where dynamic and ribbon mics prevail. Dynamic mic favorites are the Shure 57 and SM-7, Sennheiser MD-421, and e906, EV RE20 (great on smaller amps that need a low-end boost), and the Audix I5. Any ribbon mic you may have is a great choice, as ribbons and guitar amps get along nicely. The studio standards are the Royer 121, Beyer M160, and Coles 4038, although the budget-conscious Cascade FAT HEAD, Oktava ML-52, and the Golden Age and Nady R84-style ribbon mics also do a fine job.

Large diaphragm condensers set back a few feet from the raging amplifier are useful in picking up a bit of perspective on that fire-spewing beast on the studio floor. When recording George Harrison's electric guitar during Beatles sessions, a Neumann U47 or U67 was often the only mic used, simply placed a few feet in front of the amp. If you don't have these mics ($5,000 and up entry price), any LDC may work well; it's a matter of just trying them out!

Small diaphragm condensers come in last in the amp department. Naturally, if you have good room ambiance to record, a stereo setup could effectively add depth to the sound.

Typical Setups

Rarely do listeners put their ears right up to a guitar amp's speaker, although this remains the default method of mic'ing a guitar amp—jamming a Shure 57 right up against the cone. The point is to simulate the guitarist playing directly from the speakers when the mix is played back. The choice of guitar tone depends on the style of music, and how the guitar should sit in the mix. For guitar-driven songs, this may be exactly what you want. Many modern recordings err on the side of ambience, placing at least the rhythm guitars deeper into the sound of the room.

 Chapter 5 on the CD-ROM provides pictures and audio examples of some of the basic close positions and their variations. You can see and hear how small adjustments in mic position alter the tone.

One of the most important techniques for multi-mic'ing a guitar amp again requires a bit of studio trickery. You need to make sure that phase cancellations between two close mics are not hollowing out your guitar sound. To do this, you need the following:

- *A recording of white noise to be played back through the amplifier.* (This is provided in Chapter 5 on the CD-ROM, where this example is carried out, in case you don't have a noise generator.)

- *The two microphones, connected, sent two tracks and ready to be monitored.*

- *An assistant would be best, but you can move the mics yourself while wearing headphones.*

- *Prepare a way to flip the polarity of one mic, using the phase/polarity switch on a preamp or a plug-in inserted on the track.*

Here's how the trick works:

1. Place the first mic in a position you like by listening, recording a bit, and playing it back to check the sound. Mute the first mic, and do the same with the second mic.

2. Play the white noise through the amplifier at a reasonable volume. It shouldn't be too loud, because you'll need to be listening in the headphones to hear when the sound of the two mics cancel. If you have an assistant, this will be even more effective, as this person can move the second mic while you listen on the monitors.

3. While the white noise is playing through the amp, listen to both microphones set to the same level. Now flip the polarity of the second mic.

4.   In a perfect world, the two sounds would cancel out, and you'd heard silence. In reality, it's likely that the sound has become hollow or phasey sounding. The goal is now to move that second mic around a bit to get the combo of sounds to come as close to canceling out as you can.

5.   What is the point of this? Well, with one mic polarity-flipped, any sounds that would normally be reinforcing each other are now canceling out. This means that when you flip the polarity back, the sound will be all the more solid! It's naturally easier to hear the canceling-out effect than the adding effect. White noise is even at all frequencies, so it gives you a good palette of tones with which to experiment.

6.   When you've found the cancel point, you can cut the white noise. Be sure you return the polarity on the second mic to normal. Now plug in the guitar again, and let 'er rip! You should be rewarded with a full, defined guitar sound.

Tip: There is a great rule of thumb for determining whether two mics are close enough to a source to potentially have phase problems. The magic ratio is 3:1. That means that if one mic is one foot away from the sound source, the other mic should be problem-free at three times that distance, or three feet and beyond. If the first mic is two feet from the amplifier, the second should be checked for phase problems at anything fewer than six feet away. If you are using a mic and a direct signal, or two mics, where one is less than six inches from the source, it is worth checking the phase relationship in any case.

Try This

If you play back what you've recorded for the ever-fickle guitarist, and he's not impressed, have him go back into the studio, and walk about while playing until he's in that spot where it sounds, like, awesome. Then place a good large diaphragm condenser right there at head level. This should pick up the big shot's masterful guitar tone. Play that back, and blend in close mics until you have a proper mix of tone and close-mic'ed detail.

When layering guitars, alter the tone between tracks. Be sure to mark previous mic positions, or if you have the extra mics, just set up new mics. Multiple tracks will sound better if they fit like a puzzle rather than trying to just smash in the same tone over and over again. Electric guitar sounds are quite malleable, so try different effects, such as filtering or comb filtering with household objects. Putting a funnel of paper around or a toilet paper tube over a mic will create a natural filter effect. Even more pronounced are vacuum cleaner hoses, salad bowls (place it on the floor in front of the amp and mic the inside), and cardboard boxes (with a big one you can put the amp inside). Simply doubling a very straightly recorded guitar part with one of these sounds can take the tone to another level.

If you have a studio room with multiple mics already set up and connected, try listening to the various mics around the room while the guitarist plays. An interesting natural distortion effect can occur when snares rattle sympathetically with the guitar amp. The tom mics may provide an odd reverb-like sound as certain notes resonate in the drums. This sort of mic'ing by chance can be the secret sound waiting to happen.

Potential Problems

When that 8×12 Marshall stack just isn't delivering the tone that it did last weekend at the County Fairgrounds, then shrink the guitar rig. More and bigger rarely translate well into studio recordings. It's a matter of perspective; you're squeezing the sounds into stereo speakers and iPod ear buds, and they just can't handle the thunderous might of a full stack. Try a 2×12 cabinet, or even better some 10-inch speakers. When you get in close with a mic, there is plenty of low end there, and they produce a big sound at a controllable volume. If volume is an issue in your home studio, this is definitely the way to go.

Frank Zappa used to record particularly stinking guitar solos by laying a Pignose battery powered amp on its back in the middle of the studio floor, and hanging a mic from the ceiling. The parallel surfaces helped to reinforce the sound of the small speaker, making it come across on tape as a bigger amp.

Phase cancellation in a single mic can happen easily if you place a guitar amp on a hardwood floor with a mic a few feet away, in the manner of The Beatles sessions mentioned previously. See Figure 5.10.

The sound reflected off of a hard surface such as concrete or wood flooring can combine with the direct signal at the mic, causing comb filtering. If you suspect this is causing a lousy tone during your session, raise the amp up on a chair or platform made for this purpose, and lay some acoustic foam on the floor below the mic as well. This should clear up the comb filtering, giving you a more solid sound with clearer room ambience.

If you are under time pressure but cannot seem to hit "the" sound during your session, simply use a direct box, and record a copy of the dry signal as well. This can later be run out to an amplifier, processed with an amp simulator, or just used to remind guitarist what amazing part he or she was playing that day.

Mic'ing Bass Guitar/Upright Bass
Recommended Mics

Before picking out a mic, it's usually a great idea to connect a *direct box (DI)*. My personal favorite is the Countryman type 85 DI, which can be 48v phantom powered and sounds clean and clear. Since speakers act like a filter, essentially cutting off the

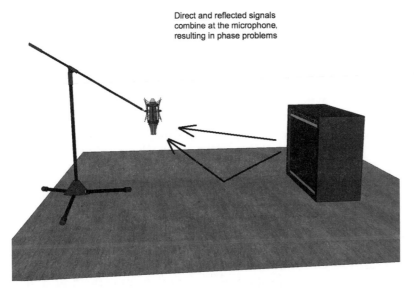

Direct and reflected signals
combine at the microphone,
resulting in phase problems

Figure 5.10   When recording near reflective surfaces, care must be taken to avoid problems caused by the combination of direct and reflected sounds at the microphone.

highest and lowest frequencies they cannot reproduce, the DI catches the "big picture" for posterity. Some upright basses are fitted with piezoelectric pickups, which allow you to use a DI on them as well.

If you are mic'ing a bass amplifier, a dynamic mic like the EV RE20, Shure Beta 52, Shure SM-7 or Sennheiser MD-421 will capture enough low end and withstand the thump and growl of an electric bass in action. Large diaphragm condensers make a nice addition to the dynamic mic sound, but again, beware of overloading them (keep the −10dB pad in mind). Try any condenser placed a few feet back from the amp, and if bleed from other instruments is not an issue (when overdubbing or running a direct line back through an amp, for instance), try omni mode to pick up some room ambience that you can blend in with the rather dry direct signal.

Large diaphragm condensers are a fine choice for upright bass, and the Neumann TLM103 and AKG 414 have worked very well for me. Blue's "Baby Bottle" mic is a great choice for stringed instruments in general. It captures a great upright bass tone, and is very affordable ($300 street price). Mic placement often makes a more obvious difference when working with upright bass, so try what you have on hand and record several tests, marking the different mic positions; then sit back, listen, and pick the position that best brings the bass's sound into focus.

Ribbon mics are outstanding on upright bass, as well as on bass amps. Small diaphragm condensers take last place here, although I've had luck mic'ing upright bass with an SDC and blending that with the direct signal from a pickup to provide some perspective.

Typical Setups

First plug the bass into the DI connecting the quarter-inch output to the amplifier, and the XLR connector to your preamp. If you're using an upright bass, and have the space to isolate an amp, you may want to run it into a small bass or guitar amp as well—the additional tone shaping this provides may help the bass's midrange cut through the mix.

If the bassist shows up with a large amp, the same rule generally applies as with a guitar; what sounds good live may not translate well to the studio. Unless the players don't care if their bass gets swallowed up by the low end of the drums, you will need to bring out the midrange character of the bass to define the finger/pick attack and general warmth of the instrument. This means leveling out that "disco smile" EQ curve on their amp and swapping those aluminum-alloy speakers for something rounder sounding.

Try This

Believe it or not, distortion is your best friend. On many well-produced songs, even the most unobtrusive bass sounds often have a portion of distortion, helping them sit well in the mix. Yep, that's the ticket! Big, clear low frequencies have some inherent weak points in commercial music.

Most typical user-end systems cannot reproduce the low frequencies that recordists, with their full 20–20kHz fidelity and monitoring can produce and hear. These frequencies are then either lost at the speaker end of the commercial system, or simply filtered out before getting there; to protect the amps and speakers.

So what's a clever engineer to do? Torture his gear. When you distort the bass signal intentionally, a couple of things happen.

■ *Distortion creates harmonics:* Depending on the gear you use to create the distortion, numerous odd and even order harmonics are generated off of the bass frequency. These higher harmonics often help the fundamental notes become more audible on consumer playback systems, and can give the impression that the fundamental tone is louder than it actually is. In cases where this fundamental frequency is below the frequency response of the playback system, these harmonics may be the only thing making that fundamental audible!

■ *Distortion compresses:* The compression is created by distortion—really hard limiting, called *clipping*. The tops of the waveform are chopped right off—allowing you to increase the level of the bass without incurring "overs," and having to turn the mix down. Harder limiting, you don't get!

■ *Distortion filters:* To a degree. If you really use a fuzz box to distort your bass, it is likely to cut off everything below 40–50Hz, which has its benefits as well. By cutting off the lowest lows, which only mutant engineers with 24-inch monitors and 2,000 watt sub-bass nuclear backup woofers can hear, you save a lot of amplifier power for the mere mortals who listen to your music. It takes approximately twice as much power to amplify a 20Hz wave to the same level as a 40Hz wave! 40Hz still sounds deep and heavy, though, so by cutting off some lower frequencies, you actually increase the power available to blast the paint off the walls. And no one (outside your studio) will even notice the difference.

Additionally, filters also create a spike in amplitude at the cutoff frequency, so a 40Hz cutoff filter would actually give a little sharp boost around 40Hz, which adds more to the punch. So, enough theory... How to do this lovely black magic? Yes, back to the torture part, you sicko.

First off, see if you have some piece of gear with both a gain (input gain) knob you can crank, as well as an output you can crank down. The best would be a Neve 1272–style preamp with an output fader, but even a channel on a Mackie-type mixer will do.

Route your bass signal to the input, and pull the output fader down. Play the bass track into the board/preamp/and so on, and crank that input until you see the gear screaming for mercy via some little red LED overload indicator. This is good—it is ready to confess. Now bring up the output fader. See if the bass sounds a little gritty around the edges. Adjust this to your taste—even a small, almost imperceptible amount will create the impression you need.

My favorite bass sound comes from my Demeter HM-1 preamp, when the red lights start to come on. But a fuzz box, distortion pedal, and many times an amplifier (even for synth bass, yes, into an amp and re-mic it!) will do the trick. I guarantee that some odd piece of gear you have lying around creates monstrous bass processing; you just need to mess around and find it.

Needless to say, it is key to this technique to listen to your mixes/sounds over a commercial system, or at least connect some consumer-style monitors in your studio. This way, you can get an idea of how this is going to translate into the real world outside of the lovely studio cocoon. By the way, this goes for drums as well. Crunching up a copy of the drums and blending them back into the mix is a tried-and-true method that started by hitting analog tape with a loud signal and has now migrated to special "analog channel" plug-ins that simulate this sound.

Potential Problems

Always check phase relationships when using multiple mics, and compare the DI phase to the mic phase using the phase flip trick. Have an assistant move the amp/upright close mic around with the polarity reversed until you hear some serious cancellations (it will sound like the low end disappears). Now flip the polarity back and you're in business.

Your control room acoustics are a serious issue when listening to bass frequencies. You may have a problem if it seems that some notes really boom and others disappear as the player performs or you listen. Try moving around the room, listening for these boomy notes and lost notes. If you don't hear the same problem in all positions, you probably have some problems with modal resonances in your control room. This is something you need to correct, and luckily it's not the most expensive or difficult thing to do. Refer back to Chapter 4, "Setting Up Your Studio."

Players with graphite-necked five-string modern-monster basses provide a unique re- cording challenge. For producers, engineers, singers, drummers, guitarists, grocery clerks, pedestrians, and other life forms the world over, there is nothing that sounds better than a Fender-style P- or Jazz bass. A P-bass fits into the mix like a beer into a cold glass. But just as double butter on your popcorn seems like a good idea at the snack stand, bassists buy basses made of carbon surrogates and sporting multiple subsonic strings. Beware! Bassists will defend their instruments, regardless of how it clogs things up. And when it comes time to mix, it's *your* problem that big, flabby bass tones are causing musical angina.

It may be that amp simulators, distortion plug-ins, filters, and exciters were designed just for this situation. Try duplicating the bass track, slapping on the plug-ins, and frantically twisting knobs hoping that something makes that funky bass show up in the mix without swallowing up all the low frequencies. Even better; keep a P-bass around your studio. Convince the bassist to double track his part with the P-bass, and then you can "balance the tonal spectrum of the bass" by turning up whichever bass sound fits the mix. Let everyone's ears make the choice.

Mic'ing Organs/Keys
Recommended Mics
As with guitar amps, dynamic mics make a dependable first choice. If you want more attack out of a Fender Rhodes or a Wurlitzer, consider using a condenser mic to pick up extra clarity. Naturally, taking a direct line from any keyboard with a ¼-inch output provides you with a clean signal and maximum flexibility. If the player uses an amp, treat it like a guitar amp as a starting point; ribbon mics are (again) a great choice.

Typical Setups
With all the excellent sounds available as plug-ins, samples, and synths, it is becoming common practice to use canned sounds for many keyboard parts. Generally, they are well recorded, sensitive to player dynamics, and the tone can be tweaked after recording by changing the patch/sample. Despite these advantages, consider running the sound through an amp at some point, and mic'ing this with one of the typical guitar amp techniques. A small tube-amp combo is a great choice, and will set your tracks apart

from the thousands of people who are using the same tone via the same synth/plug-in. By mic'ing an amp, you'll gain the tone-shaping quality of the amp, some room ambience, and perhaps a bit of welcome distortion "warmth" to give the track some character.

Try This

For more extreme stereo effects on keys, try using a spaced pair of mics. Especially in the case of a Leslie speaker with its rotating horn, a pair of dynamic mics, one on each side of the horn, will increase the stereo spread. You can regulate this later by panning the mics wider or narrower when mixing (an extremely wide stereo effect could be distracting).

To create faux-stereo from a mono keyboard sound, don't just reach for a typical mono-to-stereo plug-in (chorus/delay/reverb). Try using some real room sound and a little microphone trickery to naturally pan the keyboard in your mix soundscape. Run a copy of the keyboard track into a guitar amp. By placing one microphone close to the amplifier, and a second mic a few feet back, the second mic's signal will be ever-so-slightly delayed from the first (closer) mic. The distance of the second mic will additionally reduce high-frequency presence, and pick up more room tone, creating a sense of distance. The trick is, you then pan one mic hard left, and the other hard right.

 Although you could dial in some distortion on the amp, the point is really to put some "air" into the keyboard sound. Have a listen to the example in Chapter 5 on the CD-ROM, first dry (plug-in only) and then re-amped. By slightly rolling off high frequencies on the more distant mic, the sound appears to recede from the listener's position, giving a natural panning effect. This would work well in a mix where another instrument is panned to the opposite side; the keys will still come through in that channel, but they will take a different position in the depth of field.

Potential Problems

Good arrangement is "key," so to speak. If a guitarist and keyboard player are chording in the same octave range, the parts could compete unless there is a particular interplay they have worked out. Quite often, a keyboard part can be moved to a higher or lower octave to "get out of the way" while still being audible, and even becoming more effective.

Low notes on keyboards and especially synths can be loaded with bass frequency energy that overloads your mix in a hurry. Again, arrangement is important, and if the bass player and keyboard player run into the same octave, you can wind up with low-end mud immediately. What works in a band's 110dB barroom jams may not translate well to tape. Discuss this with the producer if you hear a lot of low-end chaos. If no one is willing to compromise, someone is going to get filtered when it comes time to mix—and it probably won't be the bassist!

Mic'ing Brass
Recommended Mics

Ribbon mics excel at brass, providing a full, natural sound. Dynamic mics like the Shure SM-7, Sennheiser 421, and EV RE20 are good choices. Just beware of the Shure 57 or similar models with potentially harsh midrange peaks. The warmth of large diaphragm condensers can be a boon to brass recordings, just be sure that your mic can handle the sound pressure of a trombone blast! Here's where the –10dB pad comes in handy. A FET or Tube condenser should be used if you have it, as the added warmth this may add to the tone and can help smooth the sound into the mix.

Typical Setups

One mic, in front of horn, press record. This is not a time to overthink things. Even if there is a full brass section playing harmonized lines for a funk record, start off by treating them as one instrument and set up just one mic. If there is a problem with the balance, have the players adjust their relative position before you start putting up individual mics. If you want a stereo sound for the ensemble, try an XY pair, and if you need more stereo spread, use a spaced pair. ORTF requires you to back away a bit, and may make the sound too ambient.

Try This

If you are working on a project wherein the producer requires absolute anal-retentive control of each individual instrument after the fact, you can put a spot mic on each piece, but please use a group mic as well. The bleed into neighboring microphones and headache resulting from trying to balance players who should have been mixing themselves during the performance may be too much to bear. The group mic could save the session and your nerves.

In fact, horn players in an ensemble should be balancing themselves well as they go—that's the point of playing together. Even if you are forced to spot mic each instrument, use the group mic as the monitoring feed. This will keep all of them on the same wavelength. It's pretty unlikely that the trombone player will want more of himself in the monitors.

After telling you that close mic'ing is probably not the way to go, here's another Beatles recording anecdote. At the specific request of John Lennon to make the horns on "Good Morning, Good Morning" sound different, engineer Geoff Emerick shoved the microphones far inside the bells of the horns. Listen to the recording if you have a copy—they sound brash and, well, different. Now you know how to get that sound if you want it!

Potential Problems

The kazoo sound is, in my humble opinion, the worst thing that can happen to recordings of brass instruments. I hear it on old Rolling Stones records, and it's a sin. I imagine that a mic with a midrange peak (Shure 57!) was used, the players were crankin', and the signal was recorded pretty hot, creating some distortion. The resulting buzzy, nasal sound sounds like kazoos. What a pity. Shy away from bright microphones, and watch your levels if the players are loud. Ribbon mics are your savior here, with their high SPL tolerance and very flat high-frequency response. If things get papery nonetheless, try backing the mic away a foot or two.

Mic'ing Strings
Recommended Mics
Ribbon mics please! Bowed strings have a rich tone, and condenser mics with their inherent resonances can bring out unwanted bow noises and string scratchiness. A smooth-sounding large diaphragm condenser is a nice choice as well. Small diaphragm condensers capture orchestral performances well, but the general close-mic application in a home studio takes third place. Dynamic mics are my last choice.

Typical Setups
In the home studio, recording strings is a challenge, because string sounds are used to generally represent orchestral performances, in a large hall with proper acoustics, and recorded from a distance.

The famous strings recording on The Beatles' "Eleanor Rigby" was done by close-mic'ing the performers, and this unorthodox method created a stir in the studio at the time. The players considered the tone harsh and unnatural, and tried to back away from the mics between takes! Producer George Martin wanted precisely this in-your-face effect, and stuck with his idea, creating a unique recording. If you are looking for a more orchestral sound, you will need a room with some diffuse ambience, or at least a dead studio to record in; you can then add hall or plate reverb later.

It is easy to create an orchestral effect with just one player, simply stack up tracks; record each individual part at least three times. The mild tuning and performance variations thicken the sound. If the instrument allows, adding low octave doubles to the part thickens it further.

A single violin (fiddle) or cello part added to a pop/rock/country recording is pretty easy to record, and the small studio ambience will probably work fine. Just use your ears, and don't get too close to the source, since a little air helps to balance and mellow the tone.

Try This
The room ambience of a home studio will make orchestral-style string parts sound unconvincing, although the natural attack of real strings is almost always better than synths or samples. Try layering some real strings for authentic attack, and adding a stereo synth/sample double to flesh out the sound.

Potential Problems

Close mic'ing a cello or violin (fiddle) may accentuate harsh string and bow sounds. If backing away from the mic leads to unwanted room ambience, try a de-esser. Tune the frequency to center on the harsh tones, and play it back in context to be sure that you're not taking away from the performance.

Bowed bass (contrabass) will really excite room resonances, leading to boomy recordings. Try positioning the contrabassist away from corners or at the very center of the room. In a pinch, try to identify the offending frequencies, and set a compressor tuned to this band to cool them out (like a de-esser for bass).

Mic'ing Percussion

Recommended Mics

Anything goes. I like to use an AKG 414, as I learned in my first big studio experiences, but you need to consider how the percussion should sit in the track. Using a Neumann U87 on a tambourine may make it so present that you just can't make it sit well in the mix (this has happened to me as well). I have since learned to pick a microphone that counterbalances the percussion's natural tone. Generally, this means using a mic that doesn't accentuate high frequencies.

I hope he won't strangle me for telling, but percussion guru Ralph MacDonald (his discography reads like a who's who of pop music) uses a Shure 57 for almost everything he records, from shakers to congas. Simpler than that, you don't get. What sounds right *is* right.

Typical Setups

Place microphone in room. Play percussion in front of microphone. Set level. Press record.

Sometimes an ambient microphone is a great addition, especially with a Latin percussion ensemble. Instruments like timbales or steel drums really benefit from some air to let the tone develop over distance. In this case, try a spaced pair of large diaphragm condensers for a bigger-than-life sound. If you want a more natural recording, a pair of small diaphragm condensers in an ORTF or baffled pair setup will create an "I was there listening" kind of sound.

Try This

Recent Tom Waits albums (since "Bone Machine"—a uniquely recorded masterpiece), which I believe heavily influenced the multi-Grammy winning Robert Plant/Alison Krauss, have used very ambient percussion sounds. By distance mic'ing shakers, tambourines, chains, castanets, and so on, the attention-grabbing high frequencies are tempered by ambience, and blend naturally into the mix. In Tom Waits' experiments, the sounds are often distorted and heavily compressed, leaving one wondering what diabolical machinery he employs to forge his songs...

Try recording odd sounds like running water, raindrops, bundles of keys shaking, raking leaves, knee slaps, and hand claps. Filter them, reverse them, trigger vocoders, and envelope filters off of the guitar or drum track. The sky's the limit. Fade these in and out during the song, and you may discover unusual depth and texture.

Potential Problems

Performance is number one. And so is the part being played. A bad percussion part or performance cannot be saved by changing microphones, EQing, or any other trick. If you are working on a project that could use good percussion and no one in the band can cut it, consider getting an expert to lay down some tracks. Nothing is more distracting than a rhythmically challenged shaker part; it can ruin the whole song. Use the foot-tap test—solo the drums and percussion, and make sure that these parts alone get a groove on. If this is not happening, something is wrong.

Again, be sure that your microphone choice is not too bright or harsh for the instrument being recorded. Beyond that, a typical amateur percussion mistake is to play too hard. You don't need to shake your teeth loose to get a good sound; it is more about consistency and dynamics. A good percussionist has incredible control and stamina. It is pretty hard to play consistent 16th notes on a tambourine!

Mic'ing Drums

You have now come to the Holy Grail of recording, getting a great drum sound. Here's the trick—it can be complicated if you choose to make it so, and the weather and position of the moon play a big role in the results. Let's consider the important elements in a great drum sound so you can reduce the influence of rain clouds and celestial bodies:

- *Quality, well-tuned drums.* This requires not only the ability to tune the drums, but also to tune them to the room and to the song. Drums have a pitch element (especially toms) and a rack tom ringing at an F# pitch over a song in C will probably sound lousy. Always keep a drum key handy in the studio, and learn at least the basics of tuning drums. The drum kit needn't be expensive, it just needs to sound good on its own. There are cheap sets that have a good crackin' sound, even if it's trashy-smashin' good. Just remember that a great recording of a lousy sound is still a lousy recording.

- *The room.* Optimally, you record in a room treated to control modal resonances and with some dispersion elements installed. Lacking the budget/permission/motivation to install these elements, you should try to isolate the drums from detrimental resonances and reflections by placing gobos around the drums. See Chapter 4, "Setting Up Your Studio" for some inexpensive ideas that will do in a pinch.

■ *Mics/preamps/levels/monitoring.* I've been over this already, and now is the time to stay organized; pick microphones carefully, label inputs well, use colored mic cables for easy troubleshooting, set levels carefully, and listen, listen, LISTEN to the results.

■ *Be critical.* The drum sound can make or break a recording. Trashy guitars, low-fi keys, and fuzzy vocals can sound great on top of a solid rhythm track, but if the drums sound like crap then the whole recording will lose face. If you need two hours of hard work just to get the mics placed right, so be it. You will learn the room in the process. Let the band know if it is the first time you're recording drums in a particular studio. This is not a lame excuse for having no engineering skills, but a simple fact that every room is different and you need time and concentration to get them what they need. If you find yourself in over your head, ask a staff engineer for some tips—there's no embarrassment in that. Usually, they'll be willing to chime in and give you some tips free of charge. A lousy recording of great-sounding drums is still a lousy recording. Sounds familiar? Yep. The pressure is on you.

Recommended Mics

When recording drums, the right mic is the one that sounds right, so whatever you have, you can make do with—I don't want to tell you to, "go out and buy mics x, y, and z or else you won't be able to record a good sound." There are, however, some standards for each application, and if you're making a list for Santa, these would be a good place to start. I break it down by application:

■ *Overheads:* Your choice of overhead mics will be highly influential on the overall drum sound. These should be bright and crisp, or do you prefer a more rounded, mellow sound (which is becoming more popular again at the start of the 21st century)? A classic overhead mic is the Coles 4038, which was the "Abbey Road" overhead mic. The drums on "Come Together" speak volumes for this mic; mellow but clear, defined, and full of impact. If you're a fan of T-Bone Burnett's production style (think "Raising Sand"), this would be a great place to start. If the 4038 is outside your budget, look into alternative inexpensive ribbon mics as overheads. You can easily score a pair for $300.

■ *Snare drum:* No brain surgery here. A Shure 57 is still the standard, although any dynamic mic will do. Some engineers like to put condenser mics on the snare drum, but I tend to keep my pricier and more delicate mics out of the direct line of fire . . . Experiment at your own peril. Experiment with the angle and placement of the mic to pick up more stick attack (aim towards the middle of the drum) or more tone and "ring" (near the edge of the drum). Some swear that the best way is to let the mic peek over the rim, "looking" across the surface of the drum. Have a listen and decide for yourself on a song-by-song basis; a quick adjustment can help the tone fit the song. See Figure 5.11.

Figure 5.11   Three typical snare drum mic positions. This graphic shows all three on the same drum, to save space. Normally you would pick your favorite sound, although you can naturally record more positions if you have the extra gear!

■ *Kick drum:* It has become standard to have a hole in the front head of the kick drum, to make it easy to place a mic close to the beater head while keeping some of the tone and resonance the front head provides. If you are recording a jazz drummer, you may come across a front head with no hole for a mic. If you need to get a mic close to the beater, you can just place one on the drummer's side, near the beater. Chances are that in this case you won't need that much "click" in the mix anyway. In some cases there is no front head, although you may still want to use a close and a distant mic. Here are some typical choices:

   ● *Up close to the front head, inside the kick:* This is a spot for a dynamic mic with a good low-end response and the ability to handle high SPLs. An AKG D12 or D112 is a classic choice, as is the EV RE20. If you're using another mic outside the head, you may just want to pick up the "impact" part of the drum sound in this position, and a Shure 57 will do well. Some engineers swear by a measurement-grade microphone in this position; something that can pick up all the way down to 20Hz. I think that in home studio situations, this may just be asking for trouble—you may just pick up the rumble of passing traffic or a jet overhead.

- *Near the outside head, or a few feet back:* The old favorite was a U47 FET or U67. If you don't happen to have one of these $7,000 mics kicking around (no pun intended), don't despair; many inexpensive large diaphragm condenser mics will do a fine job. Try an AT4033 or 4050, AKG C3000, Rode NT1A, MXL V6 or even one of their $100 condenser mics like the MXL V67. Any large diaphragm condenser helps pick up those beefy low-end tones, and lend the kick drum depth to round out the upper-midrange attack of the close mic.

- *Toms:* Sennheiser MD-421s are the standard, although any dynamic mic will do; just try them out and have a listen. If you have the spare mics, a large diaphragm condenser on each tom will pick up loads of clarity and tone. A transformerless, FET condenser will be less likely to distort under the high SPLs, otherwise engage the −10dB pad, and be sure to record a test with the drummer hitting the toms as hard as he plans to in performance; distorted tom hits sound particularly flatulent.

- *Hi-hat:* A small diaphragm condenser mic placed 6 to 8 inches away from the hi-hat will do the trick. Be sure that the mic is placed slightly above the top hat, otherwise puffs of air from the opening and closing of the hats may cause rumbles in the mic. If you don't have an extra condenser mic, use a dynamic mic; the lower sensitivity won't be a problem for pop/rock applications. If you're recording a jazz drummer, you should rely on the overhead mics for all the cymbals anyway.

- *Cymbals:* In most cases, the overheads are going to pick up all the cymbals you need. Adding spot mics on cymbals is just asking for bleed-related phase trouble. Particular cymbal effects, such as crescendos or accents in particularly soft passages, may call for a spot mic for later mix flexibility, especially if other performers are in the same room bleeding into the overheads. A small diaphragm condenser mic a foot away from the cymbal should do the trick. Just *be sure* that the cymbal will not swing and smash into your mic if hit hard!

- *"Picture of the kit" mic:* Many engineers like to place a mic 4–6 feet out in front of the kit, as if someone were standing there listening to the performance. A ribbon mic is a good choice, and the typical ribbon mic figure-eight pattern happens to work well in small rooms. It picks up the kit and room tone from the far wall without the "boxiness" of the side-to-side reflections. Perfect for the small studio!

Typical Setups

Got some free time? Good. There is no truly typical drum mic'ing setup, but there are a few tried-and-true concepts that make great starting points. Use these as a starting point, and modify to taste, knowing that you can always return to start. Your first attempts at engineering a drum sound are not likely to be an instant success, but if you follow these basics, you'll have a good starting point.

The State of the Art Until at Least the 1970s.   Two microphones. Yep, you read that right, just two—although sometimes a radical engineer would put up three, creating—are you ready for this!?—*stereo* overheads. Not much explanation required here. The overhead is positioned to pick up the whole kit. If stereo was desired, either a spaced- or XY-pair was set up. The overheads were often compressed and fed prominently into the drummer's headphones, requiring the drummer to "play to the mics." The input this provides during the performance helps the drummer find the right dynamic and tone much more easily than hearing 15 different spot microphones on every tom and cymbal. Even if you add droves of other microphones on every roto-tom and splash cymbal, try giving the drummer a simple "picture of the kit" mic to monitor with. This lets them hear what's happening and control his own tone better.

The additional microphone is used on the kick drum. The key is to manipulate the kick drum to create the desired tone. As The Beatles took interest in the recording of their records, in particular beginning with "Rubber Soul," they requested a solid thump from the kick drum. By removing the front head, stuffing a sweater inside, and placing the microphone up close, they got what they were looking for. Now you can purchase special tools designed to do just this, but you can easily (and cheaply) do this yourself. Try manipulating the kick sound by doing the following:

- *Adjusting the tension of the front head.* A loose front head can give you lots more tone and decay. Removing it makes the sound less deep, but can provide more punch.

- *Adding weights (such as sand bags) inside the drum.* If the drum is weighed down, it will move less when the beater strikes, adding more impact to the sound.

- Muffling the front head with cloths or pillows. If the front head rings too much for your taste, grab a sweater, some blankets, or such, and place them inside the drum, up against the front head. This provides a drier tone with—again—more impact.

- *Manipulating the beater.* Some beaters have a felt side and a leather/plastic/and so on side. Before reaching for the EQ knob to brighten up the kick sound, consider doing it the natural way. The harder the beater, the more click you will naturally get. Metal drummers often tape credit cards to the head where the beater hits for a serious whack! to the sound. Hell, that card was over the limit anyway . . .

Sometimes a third mic was used to pick up what they called at that time "room tone" (if the overheads were stereo, this was really a fourth mic). A room mic was placed in a spot that enhanced the depth of the kick and snare and provided a natural room ambience. If you try this simple drum mic'ing method—which is often enough to get a killer sound—experiment with mic position, or simply put up a few different room mics, and pick the one that sounds best when listening back. Figure 5.12 shows this classic drum setup.

Figure 5.12   The classic drum setup: One overhead, a kick drum mic, and (perhaps) a room mic. If you are short on tracks or mics, this technique will still get you a fine drum sound.

The Enhanced Three-Mic Method.   This is a slightly more scientific approach to using three mics, which can be attributed to the engineering work of Andy Johns (Blind Faith, Led Zeppelin, Rolling Stones, and many others) and Geoff Emerick (The Beatles). Mic 1 is placed on the kick drum, and the drum tone is manipulated as needed. The kick mic can be set back a bit, to pick up an overall picture of the kit, but be sure that the kick is thumping away prominently in this first mic.

Next, mic 2 is set up about four feet over the snare drum. This mic picks up a lot of snare (obviously) as well as hi-hat, cymbals, and high toms. Mic 3 is then placed to the right of the drummer, sometimes just behind the drummer's shoulder, sometimes toward the floor tom, even behind the floor tom, just a few inches higher than the rim.

The key is this: Mic 2 and mic 3 must be equidistant from the center of the snare. Using a piece of string, a tape measure, or whatever you have handy, just measure the distance from the center of the snare to mic 2, and swing the string around to show you where to place mic 3. See Figure 5.13.

When mic 2 is panned left and mic 3 is panned right, the snare appears perfectly in the middle of the stereo field. This also naturally pans the toms and cymbals by way of their relative distance from mics 2 and 3. This creates a solid, balanced drum sound with good stereo spread and good phase coherence. Be sure to listen to the kick drum in this balance,

Figure 5.13 By keeping the distance from each overhead to the snare drum the same, the snare appears solidly in the middle of the stereo field on playback. Cymbals and toms "pan themselves" accordingly, due to slight time differences in the arrival of the sound at the two mics.

and adjust mic 3 if necessary to bring it into focus. When the balance sounds good, blend in the front/kick mic and have a listen. If needed, a fourth mic can be added to the snare for more presence and punch.

Multi-Mic'ing for Total Control.   By the 1980s, drummers were afforded the luxury of 24 (or even 48) tracks, slick studio effects to process the sounds, and mounds of cocaine to inflate their egos. Thanks to all the close mics, EQ, compression, gate reverb, and Columbian narcotics, the resulting drum sounds were more electronic than realistic, and the trend (thankfully) began to reverse since the early 90s. Nevertheless, you may want the mix flexibility that spot mics on the toms and individual cymbals provides.

The problem with so many mics on the kit is bleed. With sound arriving at each microphone from different drums at different distances, there will be some serious phase problems. Generally, when mixing, only a few mics are faded up at once; the overheads, snare mic(s), kick mic(s), room mics (sort of a natural reverb), and perhaps a hi-hat mic. This keeps things pretty well under control in terms of phase, and then tom hits are brought up using automation when they occur. In your DAW, you can also simply trim out the silence on the tom tracks. This means removing all the bleed by simply cutting out the audio, leaving just the tom hits and their decay. If you use spot mics on the toms, I highly recommend this process.

The most significant additions to the old standard three-mic method are double mic'ing on the kick and snare, and the inclusion of stereo room mics. Here's the scoop.

The kick drum sounds significantly different a few feet in front of the drum than it does right up close to the beater head. The low frequencies are more clearly heard out in front of the kick, whereas right up close to the beater, you can pick up a lot of high-midrange attack. To get the best of both worlds, use two mics, a dynamic mic up close (which can handle the SPL) and a large diaphragm condenser out front. The close mic is often a AKG D12 (or 112), Shure Beta 52, EV RE20, or sometimes just a Shure 57. For outside the kick, a traditional favorite is the U47 FET or U67, but many of today's cheap large diaphragm condensers, like the Rode NT1A or MXL V67, work just fine. Be sure to experiment with positioning, as just a few inches movement can make the difference between a poor sound and a great one. A ribbon mic can work well, but beware the blasts of air from the kick, which can tear a ribbon to shreds. Best to stick with a condenser mic.

For additional isolation, the Hit Factory NYC kick drum mic'ing technique was to build a sort of tent/tunnel out of packing blankets the went over and out in front of the kick drum. This helps keep cymbal bleed into the ambient kick mic to a minimum, thereby improving the clarity of the image.

Mic'ing the snare drum at the snares (bottom side of the drum) in addition to the top head gives you the option of adding that crisp snare rattle into the mix. Just add another mic right up under the snare. The important point to remember is to *flip the polarity* of the bottom mic! You can just flip the polarity of the top mic (that might actually be better, if you think about how the top head moves away from the mic as the stick hits) but the important thing is to flip one of them, so that they aren't canceling each other out when blended. See Figure 5.14.

Room mics are a great addition to a drum sound if you have a large enough room to make it worthwhile. Even if the room you are in is not big enough to create any reverb, consider opening the doors to another room, or to the garage in a basement studio to let the sound reflect around and about. Place a mic in that neighboring room, get a level, and listen back. You may be surprised at the depth that a real room mic brings to the sound.

Use a spaced pair of condenser microphones for a good wide sound with no phase problems between channels. Most any mics will work fine; just try to use two of the same make and model. If you don't like what you hear at first, move the mics around a bit—it's amazing how different things can sound just a few feet away.

So, in a way, you've only added a few more mics to a basic overhead/kick/snare scenario. Even the room mics are detached from phase problems of close mic'ing, because their job is to pick up reverberant sound, and are far enough away to be past the 3:1 distance rule of thumb for microphone placement. This means that even the more complicated modern multi-mic'ing method amounts to just adding some spice to the old recipes. The only

Figure 5.14 Place an additional mic at the underside of the snare drum to pick up more "rattle" from the snares. Before you reach for the high EQ when mixing the snare, you can blend this mic to add high-frequency energy to your snare sound.

point to remember is to keep any additional spot mics silenced by automation, editing, or gating until they are needed.

It's worth noting that the problem of drums and cymbals bleeding into the spot mics on other parts of the kit has been a major challenge in getting a good drum sound since the practice of mic'ing each part of the kit got started in the 70s. The solution for getting a "bigger" sound—adding more spot mics—started to become the problem in itself! Some drummers even took to performing the basic kick/snare/hi-hat part first, then overdubbing tom fills and then cymbals in separate passes. This was studio slickness at its self-indulgent peak. This is testament to the virtues of a simpler setup, saving not only time, motivation, and tracks, but also keeping the focus on the energy and feel of the performance. With a four-mic setup in a decent-sounding room, a good drummer can provide a keeper drum track. You'll learn more about mixing techniques to squeeze maximum tone out of fewer tracks in Chapter 11, "Mixing: Balancing Art and Craft."

Try This

With a little trickery, it's possible to create a real room sound within a small studio. This is the first method I learned from engineer Peter Dennenberg during a Spin Doctors session (and is an oft-used technique among "big studio" pros). A PA system is set in the tracking room, fed with the already-recorded drum overheads (or a natural-sounding mix) of the drums. A couple of condenser or ribbon mics are then set up a ways out in the tracking room and fed back to two new tracks. It's simple really; the dry drums are played into the live room, with the additional benefit of no bleed from other instruments, and the resulting room reverb is recorded. It's simply a homemade echo chamber, and it works quite well.

The second trick, which can be combined with the first, is to turn some cardioid pattern condenser mics toward the far walls, away from the drums. The backs of the mics reject the direct sound, and a longer reverb time results. You can place hard, reflective panels (like plywood) on the far walls to "liven up" the room and create a brighter sound.

Depending on the style of music you are recording, you can use your mic preamps to further color the sound of the drums. By increasing the input gain, and reducing the output, you'll start pushing the mic preamps to the point where they begin distorting the signal. This is a favorite trick of many studio engineers, and works best on preamps with transformer outputs. Quality transformers break up nicely, adding a pleasant harmonic distortion that gives the drums an aggressive sound. If you don't have an output fader/ gain knob on your preamp, try using an external mixer. Different mixers break up differently, and for a pleasantly trashy funk drum sound, an cheap old mixer may be just the ticket. Amp and distortion plug-ins can do it too, but abusing an old mixer as an overdrive/distortion effect expands your palette of far-out sounds. Bass, keys, vocals, drums, guitars—anything can potentially benefit from that "crying for help" tone that only analog gear manages to produce.

Snare drums invite you to experiment. Spanking, thwacking, snarking good snare sounds can give an otherwise plain-jane track a well-deserved kick in the butt. There are countless stories about tracks being built around a unique snare sound. Here are some great ideas to play with:

- *Wallet taped to snare:* To increase the "thud" of a snare drum, try taping a leather wallet to the top head. You will lose brightness, but gain impact. This was a typical Andy Johns trick used during Led Zeppelin recordings. I've also had great results taming an overly bright snare by sliding an old T-shirt over it. Think of this as a natural EQ using materials instead of electronics!

- *Reamping snare sounds:* Snare drums can also gain a lot of depth and impact by being fed through an amplifier. Plug-ins work too, but a real amp also provides "moving air," which always seems to add more depth to the sound.

- *Use a different snare:* Before you go tweaking and torturing a snare sound in search of something different, consider swapping out the drum. I've had a trashy old student-kit snare laying around, and it occasionally fits the bill. Most drummers have a few snares (like guitarists have guitars . . .), so encourage them to bring some pinch-hitters to the game.

- *Re-snaring snare sounds:* Sometimes the drummer listens to the recorded take and says, "I really like the performance, but I wish I had used snare "x," instead of the one I used . . . " If you didn't think of that ahead of time, and the performance is a keeper, it's still possible to fix this! If the drummer has the snare there (or can go and get it) you can literally change the snare drum sound after the fact without a plug-in ruining the performance dynamic. Simply set up the preferred snare drum in the studio and mic it as normal. Now, create an aux send in your mixer, and send the *recorded* snare sound to that. Feed this aux send to a small monitor speaker, which should be placed face down on top of the new snare. See Figure 5.15. Now, play back the recorded snare track, loudly, into the new snare. The recorded snare hits will literally play the other snare, and you can record this "new" snare to a new track! Always keep the original performance, as you may want to blend some of that sound back into the mix.

- *Sampling noises and layering:* By using a plug-in like Drumagog or a similar sound replacer, you can create a collage of sounds to support your snare track. It's always a good idea to leave the original snare drum in there, even if it's dull sounding or somehow problematic; the dynamics of a real drummer are hard to replace without ruining the feel. The thing to do is to copy the snare track, and use a sound replacer (or manually replace each hit). If the snare drum lacks attack, try adding a snappy snare sample, handclaps, or a book being dropped on the floor. If you need more high-end, try a burst of white noise, a single shake of a shaker, or record yourself going "pshhh" into a microphone. Anything could work if blended in the right proportion!

The oft-sampled (and quoted in recording books) drum sound in Led Zeppelin's "When the Levee Breaks" was recorded with just two Beyer M160 ribbon mics. The drums were placed in the lobby of a large house, and mics stationed up the stairs that ascended from the lobby—sort of "way over" overheads. The mic signals were compressed heavily, and a tape delay was added to the sound. If you have a spare, empty Victorian-style hotel to record in, I highly recommend having a go at this technique. *Please* send me a sample!

Potential Problems
I have found that low ceilings are the most difficult problem when recording drums in a home studio. An eight-foot basement ceiling is going to reflect each drum hit right back

Figure 5.15   Finding the right snare sound for a track can make the difference between a good-sounding mix and a great one. By re-mic'ing a different snare drum using the recorded snare to "play" the new drum via a speaker, you can change the snare sound while maintaining the performance feel of the recorded track. Although the monitor is pictured above the snare (for clarity), best results are obtained by resting the monitor's speaker face-down on the top head of the snare drum.

into the overheads, causing phasing and comb filtering, or in layman's terms, "a crappy sound." To minimize this effect, the best solution is to hang a thick sound-absorbing panel above the drums. Really, it's a simple inexpensive project, so there's no excuse not to spend $25 to build and install one—just be sure it is well anchored to the ceiling! Additionally, try using a tight XY setup with cardioid mics as overheads, and keep tom and snare mics pointed downward toward the drums, so the backs of the mics help reject the unwanted reflections.

If you don't usually record drums, be sure to check out the room for rattles, squeaks, and other noises that arise when the drums are playing. Our eyes direct our attention, and it's not uncommon to watch the drummer play and hear no problems, only to later hear floorboards squeaking, glasses rattling, and rodents fleeing the scene. As always, record a few minutes of the drummer playing (both softly and at the maximum volume they'll use), both to check your levels and to listen back for problems.

Experienced studio drummers know that cymbals should be played much more gently than one might expect in order for their tone to record well. John Bonham (Led Zeppelin) was known for beating the hell out of the kick and snare, but his cymbal work was at the same time very controlled. This was key to his great tone, and it's an anecdote I like to pass along to drummers who use their cymbals for therapeutic smashing when cutting tracks.

Food for Thought: Canned Drums—Better Than the Real Thing?   In many home studios, the proper space and the required gear for recording "big studio" quality drum tracks is just not available. You can opt for one of the simpler mic'ing methods if you have the space but not the gear, but if you just can't get away with making that much noise, there are currently many options for creating top-notch drum tracks at home.

There are several approaches, and some producers choose to mix and match methods to round out a track. Consider these possibilities:

- *Electronic drum kits:* By using velocity-sensitive pads, electronic drums can capture the nuances of a drummer's performance, and save this as MIDI data, which provides all the sound-selection, and note- and tempo-editing capabilities of MIDI. As a neighbor-friendly alternative to acoustic drums, electronic drums offer drummers a familiar layout with which to "input" their tracks, although not every drummer will be comfortable on an electronic kit. They are a different axe altogether, and require practice to adjust to the feel. For those who can use them, they are a great tool for the home studio, and with the endless samples and sounds available today, you can dial in any drum sound you desire—and change sounds at any time in the production chain.

- *Drum machines and patterned-based sequencers:* Drum machines changed music, for better or for worse, at the dawn of the 80s. Some of these sounds have endured. From classic Roland TR-808 sounds reappearing consistently on anything hip-hop, to cheesy Casio keyboard beats used tongue-in-cheek on indie records, to the crunchy tones and easy programming of the Akai MPC60, these boxes provide easy operation and unmistakable tones for the drummer-less band. Although they have their limitations in terms of pattern programming and editing, it is often their simplicity and the limitations they provide that get a groove rolling along.

- *Inside the MIDI drum world:* MIDI data can be input through a keyboard, electronic drum kit, drum pads, or just by "mousing it" and programming patterns right into the MIDI window of your DAW/sequencer. Once inside, you have many options as to how to turn those notes into drum sounds. Programs like FXpansion's BFD, Reason (with numerous Refill sound banks), Big Fish Audio's Addictive Drums, and a growing group of peers offer easy-to-use interfaces for putting together virtual drum kits (really just a way of organizing samples) which are then "played" via MIDI. The samples often allow you to switch drum mics, adjust mic placement, and process the sounds with various effects. The samples are now so good, and there are so many parameters to mess with that it's almost too much. Provided you can create the MIDI data in a groovy way, you'll have drum tracks that probably sound too good for your other home-baked tracks!

Because MIDI allows for easy interconnection, you can naturally use any MIDI sound module for your drum sounds. However, with so many host-based programs available, the additional cost of a hardware MIDI sound module is outdating them quickly. The major advantage of a dedicated sound module for MIDI drums is to free up your computer for other tasks while recording—the drum module can be used just for the drummer's monitoring, while you record just the CPU-friendly MIDI data. You can score a decent-sounding module from Yamaha, Alesis, or Roland for next to nothing if you browse the normal used gear sites and shops.

- ■ *Loops:* Even drum loops are experiencing a renaissance, as clever programmers devise new ways of making the otherwise inflexible loop into a pitch- and tempo-flexible production tool. Just a few years ago, loops were limited in their usefulness as far as how much you could change the tempo and pitch before they sounded too "processed" and drifted out of the groove. Reason's .rex format is a perfect example of how a little tricky thinking has revolutionized loops. A loop is just a sample of a musical phrase that's played over and over again. By splitting the loop even more—into individual segments, beat for beat—the tempo and pitch can be stretched to fit your production without audible artifacts. At least to an extent. By the time you stretch a .rex loop so far as to hear the detrimental effects of the processing, you're probably way outside of a usable range for that loop anyway. If you happen to have a library of CD drum loops collecting dust, consider bringing them back to life by converting them into .rex loops; I certainly have plenty to do on the next rainy day . . .

- ■ *Sample replacers:* Plug-ins such as Drumagog can replace any percussive sound you record with any other sample you choose. Most often, engineers use these types of programs to either replace or support recorded drum tracks with more samples. The concept is simple; you tell the plug-in to analyze a track, and play the type of sample you load into the plug-in whenever the trigger track reaches a certain level. You can even set multi-samples, so that the plug-in will replace a harder-hit trigger sound with a harder-hit sample, allowing for a more realistic dynamic in the replaced sounds. The cool thing is, as long as the sound is percussive, you can replace literally *any* percussive sound with a drum sample. In this way, you could stomp your foot and clap your hands on your lap, imitating kick and snare. Put one mic on your foot, one on your knees, and get slap-happy. Then run the plug-in to replace the foot with a studio-quality, multi-sampled kick drum, and the knee slaps with snare. It sounds far out, but it works; once the samples are in, it doesn't matter how they were triggered—if it sounds good, it *is* good.

Currently, there's a trend in some circles toward simpler drum recordings; the old "three-mic" techniques have an immediacy and simplicity that serves the impact of the music. The White Stripes (and their many imitators/inspired-bys) are a great example of artists trading slick studio sounds for raw impact.

Mic'ing Unusual Instruments
Recommended Mics
When recording an instrument you have never dealt with before, it's good to start with a large diaphragm condenser mic that you are familiar with. Try to find a good position just by listening, placing your trustworthy mic where the instrument sounds good, and making a test recording. For percussive instruments, grab a Sennheiser 421 if you have it; that's a trustworthy dynamic mic with a wider frequency response than the old Shure 57.

Typical Setups
Sitar, accordion, concertina, penny whistle, bodhrán, Celtic harp, jaw harp, bagpipes, baglama, balalaika—there will be no "typical" setup, but you can use your ears to guide you. For instance, bagpipes generally do not sound best close up. I recommend trying a distant mic'ed approach, moving the mic closer until you find the right blend of direct and ambient sound. Stringed instruments like sitar, balalaika, kamanche, saz, and so on may call for a guitar-like setup. For bodhrán (Irish folk single-headed drum), you probably should not close-mic it either, as the tone of the drum "opens up" over a distance. Accordion and concertina also require a bit of "air" to allow the tone to mellow somewhat; close mics may pick up too many motion sounds as the player works the bellows!

Try This
If the instrument is not featured in the track, but rather a supporting overdub, be cautious not to use a mic that brings the instrument "forward" in the mix (a bright mic, for instance). Ribbon mics with their mellow frequency response are likely to pick up the instrument as you hear it in the studio, just beware, again of instruments that may create puffs of air that could harm a ribbon mic. If all else fails, grab a simple dynamic mic and see what happens . . . they have saved the day more times than you can imagine!

Potential Problems
The "unusual" instrument may often be some sort of traditional instrument, and the performer will usually expect an accurate recording, rather than an experimental approach. In this case be sure to invite the performer into the control room for a listen, and make sure they are happy with the tone. Close-mic'ing can sound unnatural on many traditional instruments, so if you have an unhappy performer, try moving the mic back a few feet, or placing a mic in the "player perspective" that you've used before—a foot or

so over their head, as if the mic were listening with them. Alternately, try one of the stereo methods covered earlier. Ask the performer how far away the audience might typically be, and place the stereo array into the position where their audience would normally be. This will give a very natural impression of the performance. Use sound absorbers behind the performer or the microphones if you hear too much ambience.

Final Thoughts

That's an awful lot of information for the budding home engineer to digest! Just remember to use what you have available, and not to obsess about a special microphone that might make or break your recordings. Listen to the source, and experiment with mic placement to find a sound that works. In many cases, the placement of the mic will make as much a difference (if not more) than switching mics.

Even with just a couple of mics, you should be able to follow your ears and cut some fine-sounding tracks, but we all hit a brick wall from time to time. Necessity may be the mother of invention, but frustration is her bratty little brother! If nothing is working, consider changing the sound source. Perhaps the guitar you are trying to record just doesn't fit the song. In this case the microphone won't help; you need to remember the priorities. Song first, performance next, then sound source, and *then* microphone technique. It's fun to mess around with microphones, but the wise engineer keeps a good overview of the whole process.

6 Signal Processing Toys: EQ

So far, you've learned about the recording process, and how important an artistic concept is to the process. You've learned about the qualities of microphones, and techniques for using them, as well as the importance of good acoustics in your recording space. Now you can move on to playing with all those great toys you haven't touched yet—equalizers, compressors, delays, and reverbs, YEAH!

Before you go and start twisting knobs, let's run down the all-important list of priorities, and see where you are:

- Inspired song/performance—Check!

- Well-tuned/good-sounding instruments—Check!

- Proper room acoustics—Check!

- A reliable monitoring system—Check!

- Good engineering technique (mic technique)—Check!

- Signal chain/processing—Let's dig in!

You've done it all correctly so far, so let's not ruin all the hard work by twisting the tracks out of shape with heavy-handed processing. The key to using EQ, compression, and more, is to apply these effects to *support* an already solid recording. If you've jumped ahead to this point because you want to "fix" a recording that you are not happy with, you're out of luck. It is by far an exception, and not the rule, when some sort of plug-in or effect makes a bad recording sound good (although tuning plug-ins like Auto-Tune do often save the day). If you're basically unhappy with a recording, the place to fix it is in the performance, room acoustics, or engineering technique.

Example: The recording of a bass guitar amplifier sounds "boomy" on certain notes, so you reach for the EQ. You find the "boomy" note frequency, and cut it way back. What you wind up with is a bass sound with a big hole in it; by cutting so much, the natural tone of the instrument doesn't come through, and the recording sounds weak.

127

The solution is in good engineering practice. You should always record a direct line from the bass, in case of such situations. Fix the sound at the source—pick a good mic for the task, position it right, listen for "boominess," and place sound absorbers to control the acoustics. This takes a bit more effort than just grabbing the EQ knob and twisting, but the EQ "solution" doesn't really help you! The basic tone needs to be there—time spent on recording great basic tracks pays back a hundred fold when it comes time to mix.

The best engineers in the business don't reach for the EQ if the sound isn't quite right—they move the mic, or try a different mic. If the snare drum isn't bright enough, swap out the snare drum, and maybe add a mic underneath, near the snares. Guitar sound too boomy? Move the mic, try a different mic, or try a different guitar.

When the engineering side is in shape, signal processing can make good-sounding signals from your carefully placed mics into great-sounding recorded tracks.

Equalization Techniques

Consider for a moment what the word "equalization" implies. Equalize means "to make equal," or "to make uniform." Although it has become popular to heavily filter sounds to create ear-catching tracks, the original goal of audio equalization (EQ) was to correct irregularities in a sound. EQ was used very conservatively until the 60s, when artists and producers started cranking up the high end for "sparkle" on drum overheads and guitars, filtering vocals to create odd effects, and boosting the lows on drums and bass to get things shaking.

Analog tape tended to lose high-frequency detail, so boosting treble EQ on the way to tape was a common way to keep tracks sounding clear. By adding an exaggerated amount of high EQ on the way to tape, engineers could turn down the treble on playback with the added benefit of turning down the tape hiss! A clever concept.

As these EQ habits spread throughout the music industry, manufacturers of microphones, guitars, amplifiers, and even drums and cymbals began to favor bright-sounding gear. Over time, this EQ boost became built into much of the gear used.

Now, most people record on digital media, which no longer has the characteristic "softening" of high frequencies that analog tape did. So with all these bright-sounding mics, brash guitar amps, and zing-y drum cymbals being recorded onto very accurate digital media, it's no wonder that many listeners think that "digital is harsh." The truth is, digital is just honest. Garbage in, garbage out, as they say. Well, maybe not garbage, but what can be done?

Consider for a moment the recent gear trends—analog channel plug-ins, tape simulators, tube gear with a "vintage" tone. What is it that we're all looking for? I'd venture to say: "Less treble!"

In fact, what many popular toys offer is a way to "cool out" that digital/treble ear-burn. In addition to that, there's often a dash of harmonic distortion, reminding recordists again of the irregularities of analog tape.

So, to use EQ wisely, home recordists using digital interfaces and hard drives need to come up with a new approach to EQing tracks. Just about any book on recording technique gives standard EQ advice such as:

- "Boost 12kHz on the drum overheads for more clarity"

- "Add a few dB at 3–5kHz to increase the intelligibility of a vocal"

- "The snap of the kick drum can be brought out with a narrow boost around 2kHz"

These kind of tips are essentially outdated, although they are still widely touted across the Internet and in many publications as "The Way to Use EQ." What happens with this approach is that a race is started—the cymbals sizzle, the vocal cuts, the kick drum snaps, and then, in comparison, the guitars sound dull, and the bass doesn't cut through. So the confused engineer reaches for more EQ, trying to make everything uniform again. The result is a harsh mix.

EQ Overview

Let's approach EQ in a different way. The analog days are over, and you need to adjust studio methods to the current technology. The following techniques are used by many "pro" engineers, and don't cause the ear-grating imbalances that those old, typical rules of thumb can lead to.

There are many types of EQ, so an overview is in order:

- *Graphic EQ:* You know what this type of EQ device looks like if you've messed around with home stereos; there are sliders for each frequency, and it's more often than not set to the "disco smile" curve, as shown in Figure 6.1.

  This is not to say that it is a bad EQ design; there are certainly quality units available, and the visual interface makes them very intuitive to use. You'll often see them used in live sound applications, for easy EQ control on monitoring devices, and on guitar amps and pedals. They invite you to just grab a fader, yank, and hear the results!

- *Shelving EQ:* Commonly seen as the typical "low" and "high" EQ knobs on a radio, a simple guitar amp, and so on. The frequency is fixed, and turning the knob simply increases or decreases the gain from that frequency on down (for "low") or from that frequency on up (for "high"). With a shelving filter, the frequencies below (or above, for a high shelf) are cut or boosted by the same amount.

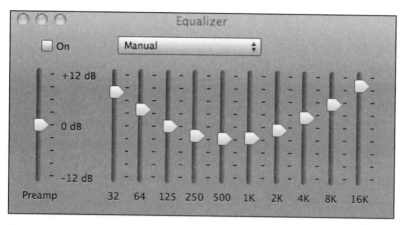

Figure 6.1 Say cheese! iTunes has become the modern "home stereo" and many of the preset EQ curves feature boosted bass and treble—the classic "disco smile."

With modern shelving EQs and EQ plug-ins, you can also vary the frequency at which the "shelf" starts, allowing you to tailor the filter to your needs.

- *Filters:* Easy to understand, filters always let certain frequencies pass through, while attenuating everything else. So, a "high-pass filter" (HPF) allows the high frequencies to pass above the chosen HP frequency, while cutting everything below that. Same goes for "low-pass" (LP) and "band-pass" (BP). For BP filters, you will set two frequencies, one high, and one low, and the band in between is allowed to pass. Filters do not equally reduce frequencies below (or above) their corner frequencies, but rather cause them to fall off sharply. The "slope" of a filter describes this, and is rated in dB/octave. So a 12dB/octave low-pass filter set at 400Hz would reduce 200Hz by 12dB, 100Hz by 24dB, 50Hz by 36dB, and so on.

  Filters with high slopes tend to resonate at the corner frequencies, creating a small peak in frequency response. This can be used to good effect when processing a signal, as it makes the sound stand out in the mix. Classic EQ and filter devices by Pultec are prized for their particularly musical resonances.

- *Parametric EQ:* Allows users to adjust three *parameters*—the center frequency, the bandwidth (often the value Q is used, which is related to bandwidth), and the gain of the chosen frequency. For adjusting the frequency spectrum of individual tracks, this is your most powerful tool, and very important to understand.

Being able to choose the frequency you want to adjust is the first obvious benefit to a parametric EQ. This feature allows you to hone in on a problem area and adjust it; something that is not possible with other types of EQ. Figure 6.2 shows the UAD 1081 EQ plug-in, based on the Neve 1081 module, which has become famous for its tone and

Figure 6.2 The UAD 1081 EQ design (here as a plug-in) offers two bandwidth settings, instead of variable adjustment—narrow and wide. To save space, the frequency is selected by the outer ring of the knob, and the gain on the protruding inside knob.

flexibility. Originally built to be the preamp and EQ section of a larger console, many of these classic preamp/EQs have been rack mounted, and have found their way into studios across the world.

Modern-Day EQ Advice

The Q or "bandwidth" setting allows you to adjust by how much neighboring frequencies are affected. Q and "bandwidth" are related by fairly complex mathematical equations, and for the purposes of this book, it suffices to say that this is the adjustment for "how much are neighboring frequencies affected," like the slope of a filter. A wide bandwidth, or low Q value affects a wider swath of neighboring frequencies, resulting in a gradual sloping off of the affected frequencies. High Q values are equivalent to a narrower bandwidth, and therefore a steeper drop off of affected frequencies past the center frequency.

With hardware EQ, you'll need to use your ears to make adjustments, which is the best practice anyway! With many plug-in EQs, a graphic depiction shows you what is happening, although this should *always* be secondary to what you hear.

Figures 6.3 and 6.4 compare different settings on a parametric EQ.

Let's consider the effect bandwidth has on what you'll hear when adjusting the EQ on a track. With a very wide bandwidth (low Q), you affect neighboring frequencies much more when boosting or cutting. Let's say you have a guitar track that is overall very midrange-heavy or "boxy," as some might say. By centering the frequency at, let's say 700Hz, and cutting about 3dB with a *wide* bandwidth, the sound loses a lot of midrange overall. In this case, it's almost like boosting the highs and lows (the "disco smile" effect), only achieved by cutting the mids, as shown in Figure 6.5.

Imagine that you used a Shure SM-57 to record a guitar amplifier, and although the performance was great, that typical 5kHz "bump" in the mic is harsh on your ears. Is this a lost cause? It would have been preferable to choose a mic without that inherent upper-mid boost, but no worries; a touch of EQ can save the day. See Figure 6.6.

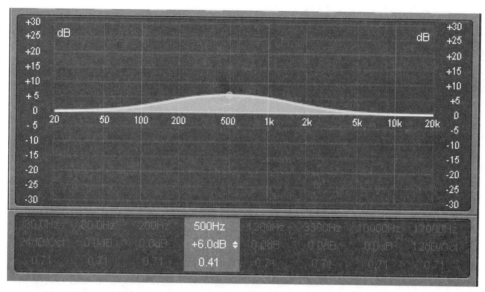

Figure 6.3   The excellent channel EQ in Logic. Pictured is a parametric EQ set to boost 500Hz by 6dB, with a Q of 0.4. That's a very wide bandwidth; about three octaves.

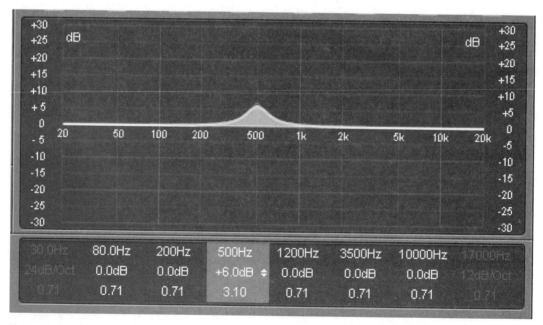

Figure 6.4   This image shows a very narrow boost at 500Hz by 6dB. The Q is set to about 3, which gives you a bandwidth of less than half an octave.

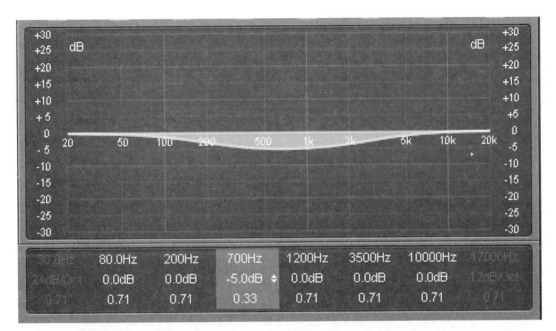

Figure 6.5   A wide bandwidth cut centered at 700Hz. One benefit of parametric EQ is the ability to make wide adjustments to the overall frequency content of a signal.

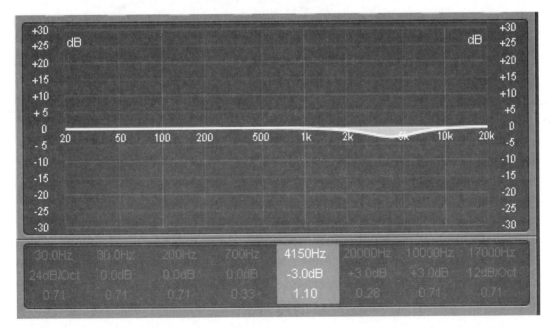

Figure 6.6   A moderate bandwidth cut of 3dB, centered at about 4kHz. This frequency range can sound harsh in some cases. If the sound source already has "bite" in that range, and the microphone accentuated this, some corrective EQ can help. By narrowing the bandwidth, you can focus only on the problem area, and maintain tonal balance above and below the EQ's effective range.

Very narrow bandwidth (high Q) settings can also be very helpful. There is a technique many engineers use to identify problem frequencies when applying EQ, sometimes referred to as the "boost and sweep" method.

The concept is brilliant—it's like the audio version of a magnifying glass. By setting the bandwidth fairly narrow and boosting by 6dB (or a bit more), you can use the frequency parameter to sweep through the audio spectrum. If there's an undesirable frequency that's sticking out, it becomes glaringly obvious when this "magnifying glass" passes over it. See Figure 6.7.

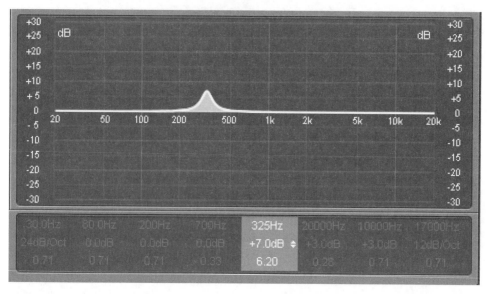

Figure 6.7   Looking for trouble... Logic's on-board EQ lets you dial in a very narrow bandwidth for easy problem hunting.

Having found the unpleasant frequency, you can now cut it back. Generally, you widen the bandwidth when cutting, as narrow bandwidth boosts and cuts can sound harsh or unnatural in themselves. There are cases, however, where a narrow cut is needed. When you find the naughty frequency, cut it and adjust the bandwidth while listening. Be sure to listen to the EQed track in context, as well, because the way it fits in with the rest of the mix is key. See Figure 6.8.

So, wide bandwidth adjustments affect a broader portion of the track's sound, but tend to sound "smoother" and more natural. Conversely, narrow bandwidth adjustments can be used to affect a smaller range of the audio spectrum, but can potentially sound unnatural in themselves. If you need a rule of thumb, err on the side of a wide adjustment, especially when boosting (increasing gain on the EQ); it just sounds smoother.

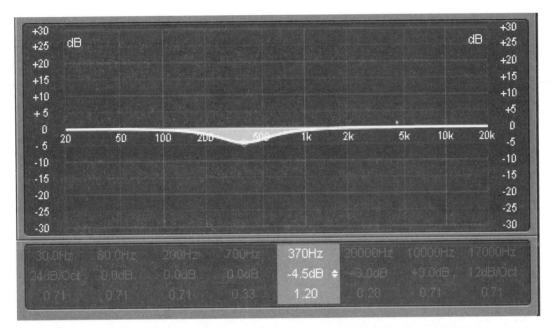

Figure 6.8   Having found the offending frequency, you then broaden the bandwidth to make the adjustment sound more natural, and cut the frequency. This is perhaps the most important EQ technique you have at your disposal.

INSIDE THE BOXES
Understanding EQ, Harmonics, and Tone   While on the subject of EQ, this is a good time to get a feel for the raw materials of which sounds are made. In theory, frequencies can be pure tones with predictable characteristics. For example, a tuning fork rings at 440Hz (the A above middle C), and the low E-string of a guitar is tuned to about 82Hz.

In practice, however, these sounds are complex waveforms made up of not just their basic note ("fundamental frequency," for you Propellerheads), they are a combination of this basic tone and numerous other *harmonics,* and the whole she-bang is then affected by the physical shape of the instrument and its inherent resonances. And then the acoustics of the performance space come into play as well!

Because the guitar is essentially such a simple instrument in its construction—a taught wire fastened to a piece of wood—it makes a great example for the complexity of sound.

Let's say that you connect an electric guitar to an overdriven amplifier, and pluck the low E-string. As noted, the fundamental frequency of that note is about 82Hz. What comes out of that amplifier, however, is a lot more than a pure 82Hz tone! Especially since the amp is overdriven, you're going to hear a combination of the fundamental tone, its first, second, third, fourth, (and so on) harmonics, the interaction of the amp overdrive accentuating these harmonics, the physical interaction of all this with the amp's cabinet and speaker, and the influence of the room acoustics on everything.

Harmonics are multiples of the basic frequency (f). So, the first harmonic (1 × f) is actually the note itself—the fundamental. The *second* harmonic is the *first* overtone (2 × f); that terminology can be a bit confusing, but the idea is simple.

What happens, as you go continue up the harmonics (2 × f, 3 × f, 4 × f, 5 × f, and so on), is that you eventually run into all the notes in the musical system; yes, all 12 tones. Granted, they are not all audible as separate tones, but they are, in some small amount, blended in there. The blend of these different overtones, along with the sound-shaping qualities of the instrument's physical build, produces the overall tone of the sound called *timbre* (pronounced "tamber").

This is why, despite the pitch of that guitar string being 82Hz, tweaking the high EQ on a recording of that note will affect the sound. There are high overtones (overtones equals harmonics), making up the unique timbre, which are affected by EQ adjustments no matter how you turn those knobs!

To give an example of when a very narrow cut may be useful, let's consider a guitar solo, played high up on the fretboard. Sometimes a particular note jumps out in an unpleasant way; a combination of guitar and amp resonances that just cuts off the top of your head. In this case, a very narrow EQ cut could focus right in on that note and pull the volume down. With a bandwidth of $\frac{1}{12}$th of an octave (there are 12 notes in an octave, so $\frac{1}{12}$th of an octave is one note), you could theoretically EQ one note out of a performance! In practice, sound is far too complex for this to work perfectly, but you get the idea—a narrow bandwidth (high Q) cut could do the trick. There are a few ways to do this, and I'll revisit this concept later when I get to *de-essing*. For now, let's just use EQ:

- You can identify the problem frequency with the "boost and sweep" method, but with high notes, this can be unpleasant to listen to. Alternately, you could "cut and sweep," listening to hear when the harsh frequency disappears. (It's usually easier to hear boosts than cuts, just like you're more likely to notice something new rather than something missing—a psychology thing.)

- Adjust your EQ gain and bandwidth to taste.

- Next, so you don't have to leave this unnatural-sounding cut engaged during the entire track, automate the EQ in your DAW to reduce the gain only during the problem notes. Problem solved.

In general, narrow cuts/boosts are best for problem solving or special effects, and, as you'll learn, there are often better solutions than EQ cuts for typical problems like sibilance or "peaky" resonances in your tracks.

Equalization Tips

Admittedly, not every track is going to have a proper EQ balance just from good engineering technique. It's pretty rare that a track makes its way into the mix completely dry; some kind of sauce usually tops off a track to make it more appetizing, and the right touch of EQ will make a good track sound fine.

But instead of just rattling off a list of instruments and explaining how to EQ them (which is futile, as there are as many unique tones as there are good players), this section investigates the whole audio spectrum, and explains how the impression the different frequency ranges give to different instruments. Have a look at this *very* subjective, poetic graph of the audible frequency spectrum shown in Figure 6.9. Above the line are "positive" descriptions of that region of sound, below are "negative" descriptions. Sometimes a particular frequency band sounds good in a track, sometimes bad. For instance, a vocal with too much 700Hz could sound boxy, but a guitar part with a prominent boost there may sound funky, chunky good.

Figure 6.9   The author's garbled attempt at communicating the impression of sound. Please bear with him, he is trying to express himself, and has a contractual right to do so. He gets snippy when people try to stop him.

There are some pretty typical EQ moves that everyone should be aware of as well, and you'll also learn a few of these, especially in relation to drum EQ, which is the apprentice engineer's most difficult task.

Let's start at the bottom of the frequency range, and move up in segments at which the impression of the sound changes in a particular way when altering that range's balance. These are guidelines—rules of thumb—as is all written advice related to audio! You have to use your ears and trust them. Keep in mind that an acoustically treated control room and a few different sets of monitors are essential to being able to make good EQ, effects processing, and mixing decisions.

20Hz–50Hz: Way WAY Down at the Bottom

Truth be told, many of you with smaller monitors and/or no sub-woofer won't be able to hear much in this range. This can be a problem, because what you can't hear is out of your control. If you can't hear 40Hz, and there is a lot of 40Hz energy in a track (for example, in a low synth pad, bass drum, or even the physical rumbling of a mic stand being shaken), that 40Hz energy may be eating up headroom in your mix and creating a big heap of lousiness that you won't hear until you listen to the mix on a system that reproduces these tones.

Fact: The low E-string of a bass guitar has a fundamental frequency of about 40Hz.

"I can hear that!" you say, "Make it fatter! Boost 40Hz!"

Remember harmonics and timbre! You may very well hear 40Hz, but you're hearing a lot more tone from upper harmonics. Even 80Hz and 160Hz (1st and 2nd overtones) are going to be a better bet to crank up to get a fatter bass sound. If you load up on 40Hz, you could also bring up woofer-destroying thumps from finger noise or a string hitting the pickup. You'll also make your compressor kick in harder, crushing the track's dynamic and ruining the "impact" that you intended to crank up. (You'll learn about compressors and low frequencies in the next section.)

Tip: Five-string basses, with a low B-string, need help to even appear in the mix; the low B note has a fundamental frequency of about 30Hz. The wavelength may not even fit in your studio. Please add *some kind of distortion* to this bass so you can at least hear the harmonics!

Because the LOW low bass frequency band can be such a mix killer, especially for home-studio engineers, who most often have compromised monitoring situations, you need to take some precautions:

- *Rumbles:* When tracking, take precautions against mic "rumble." A typical problem is the foot-tapping, limb flailing, P-popping singer. If possible, use a "spider web" type microphone shock mount, as shown in Figure 6.10. This does a pretty good job of decoupling the mic from the stand, and therefore the floor, keeping low-frequency vibrations from rumbling in the mic diaphragm.

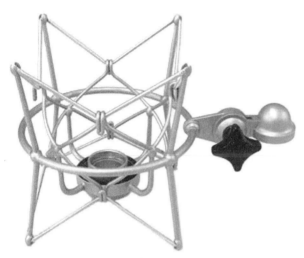

Figure 6.10   In case you are unsure what a shock mount looks like, this is a typical, high-quality mount from Neumann. Better quality shock mounts use very flexible elastic. The stiffer the shock mount's "web" material is, the more it will transfer vibrations, so a good quality mount is worth the investment.

- *Pops:* Likewise, protect the mic from "popped" consonants (from singers) and other bursts of air. A pop filter is essential for vocal tracking, and if things get heavy, engaging the low-cut filter on the mic is also an option. Drums can send bursts of air flying some distance across a room. The hole cut in the outside head of a kick drum can create heavy rumbles in the ambient kick mic, so position it carefully. In these cases, always solo the track and have a listen. Also look at the track's meter; a burst of air may cause a low rumble that you don't hear, but pushes the tracks meter (which hears everything!) into the red.

- *Heavy-handed boosts:* When using EQ to boost the lows on a kick drum, bass part, and so on, be skeptical about big boosts (6dB +) from which you don't hear an immediate difference. In this case, you may be boosting some LOW lows that your monitors don't reproduce. Try boosting harmonics of these lows, which you and other mortals can hear.

For example, if you want a big, fat bass drum, you may want to boost 50Hz. However, many listeners won't hear much 50Hz energy in their playback. Instead of cranking up 50Hz by 12dB, try adding just a few dB at 50Hz, a few at 100Hz, and a little bump at 200Hz. This will be much more audible, and translate better to all playback systems. Likewise, you could process the kick with a bit of distortion, which naturally creates higher harmonics, thus warming and fattening the sound in a natural way. If an amp plug-in cuts too much low end off the kick drum, set up the plug-in on an auxiliary send, and use the amp plug-in as you would a reverb, sending a copy of the signal to the plug-in, and then blending as much of the sound as you like.

■ *Uninvited energy:* There are many instruments that have no business adding energy in the 20–50Hz range to the mix. Anything you record live using a condenser microphone may have unnecessary low-frequency energy in it, and many of these instruments don't even naturally create musical information in that range.

For example, guitars (and any non-bass stringed instruments), keyboard parts with no low-range notes played, high percussion (shakers, maracas, chimes), and high-range vocals most likely have no useable 20–50Hz info in them. You can safely filter off the lows from the track, losing only the upstairs neighbor's size 12s, the garbage truck's rumbles, and the turbines of a passing 767. Again, always solo the track and listen. If the track loses a certain "gut feeling" when the lows are filtered, turn the filter off unless you hear a particular problem. Less is more.

There are anecdotes about engineers creating the "best kick drum sound" by cranking up the lowest frequency they can dial in. So why shouldn't you do this? Hey, nothing's forbidden; give it a try if you want. Let me just say two things: 1) The mix engineer had much better speakers than you do, and could hear what he was doing. 2) The mastering engineer's first adjustment was probably to "turn down all that stupid 20Hz crap in the mix."

In general, be cautious with the very lowest of low frequencies; they're more likely to cause problems than to elevate your mix into awesomeness.

60Hz–100Hz: Effective Bass

Here's the range that can help you can add some beef to your mixes (for you vegetarians, think of it as seitan). In any case, it's the more environmentally friendly beef, and you'll probably be able to hear most of this range clearly, even on modest monitors.

The 60Hz neighborhood is a solid, deep-sounding bass range that no kick drum should be without. A little 60Hz boost is usually in order, just don't overdo it. Otherwise, when it comes time to master, some 60Hz may have to be cut out, and that will have to be cut from the whole mix, actually *stealing* phatness. Lame.

The fundamental bass notes of your song fall between 60Hz and 100Hz, so this is the "base" of your song. You want to be sure that you keep this range clearly audible. This means careful listening, not just liberally boosting this range on every track.

(As noted, the bass guitar's low E-string is about 40Hz, although the more audible root will be at 80Hz. Again, those very low frequencies may not be audible on all playback systems.)

Let's say that you're working on a pop song in the key of A. The bass guitar's low A-string is tuned to 55Hz, the A-string on the acoustic guitar to 110Hz, and the drummer has his kick tuned in right, pumping away in the 55Hz range. There's also a keyboard part, supporting the arrangement with chordal parts.

To keep things clear in the low end, you may want to cut a little out of the guitar and keys in the 55Hz range (again, you should be using a medium-wide bandwidth to keep things musical sounding). Why? If their fundamental notes are above this range, why would you cut at all? Isn't that simply "below their range"? Well, the very act of strumming the acoustic guitar creates percussive sounds that may have some near-60Hz energy, and that keyboard patch may have some synth-y low frequencies in there; and those sustaining chords rob "punch" from the bass and kick drum.

I'm not talking about a lot of EQ here—you don't want to rob the acoustic guitar and keys of "body"—but you want to allow the bass and kick to control that low-end motion. This is a subtle thing, especially if you've taken care to record the basic tracks right in the first place.

Food for Thought: Thinking About Arrangement   Music depends on contrasts to maintain the listener's interest. You may want to pick up a copy of Eno's "Oblique Strategies" (as software, or as actual cards) to help you think critically about things such as the contrasts in music that keep attention and interest. Incidentally, I just pulled a card, and it read, "Is something missing?". In that case, let me add a few notes which you may want to copy onto a "cheat sheet" and keep handy at your mix position when thinking about contrast in terms of arrangement:

- *Low vs. High.* Can the part be played an octave higher? Or lower? Or doubled?

- *Loud vs. Soft.* Is the part a surprise when it enters? Or has everything been cranked to the same boring level?

- *Thick vs. Thin.* Sometimes triple-tracking a part helps; sometimes it takes away impact. Add or reduce and be critical. Maybe a thin snare sound is more effective, maybe the background vocal should sound like a choir to get the point across.

- *More vs. Less.* Similar to thick/thin, but over time. Consider breaking the arrangement down for effect. Remove the bass, percussion, and keys. Enjoy the "air" this creates. Then bring it all back in for a big sledgehammer to the forehead.

- *Repetition vs. Variation.* Parts that repeat tend to fall into the background, whereas new rhythmic and melodic elements jump out at the listener. Think the drums aren't loud enough? Consider that they may simply be too repetitive. Vocal harmony parts are especially sensitive to this; if a harmony has been performed throughout the song, consider where it could be repeated or varied to keep it from stagnating.

- *Simple vs. Complex.* This speaks for itself, and varies depending on musical style. What is complex in Pop may still be simple in Jazz. What serves the song best?

- *Chaotic vs. Ordered.* Order communicates structure, chaos adds variation and interest. Harmony, melody, rhythm, timbre, dynamic—all can be made more ordered or chaotic. Experiment.

To get back on track in regards to EQ, arrangement also plays a role in how much energy there is in the low frequency range. Those of you recording "heavy" music (be it metal, dance, or experimental) should be aware that contrast keeps things "heavy." Simply piling on detuned guitars, low-end heavy drum samples, five-string bass, and using EQ and compression to heap on the lows and crank the levels may actually make your music sound "small" in the end.

Think of it this way: If the guitarist cranks out more low end than the drummer's kick drum, which will sound "bigger"? It will be relative. And when it comes time to mix and master, the low end must be put into balance with the mids and highs, or else the mix will not sound full—it will just be muddy.

The best engineers for "heavy" music styles know that the key to heaviness hides in the midrange. The definition of each instrument's physical size and shape is perceived here, and the tones must complement one another to come together into a truly "heavy" onslaught of sound. So, look beyond the low end for heaviness, and keep the "low lows" under control for the sake of the mix.

100Hz–250Hz: Thickness

This range is the home of the first overtones of all the bass notes in your arrangement. As long as the instrument itself has a balanced timbre, its energy in this range is likely to be in order, and you won't want to adjust too much here. If you happen to have a bass that sounds "woofy" (too thick) you should do some "boost and sweep" to find that thickness, and do a broadband cut of a few dB.

Snare drums have a lot of energy in this range, as do congas and other large percussion instruments. If you want your snare to sound larger, this is the magnifying glass range! Do a "boost and sweep" to find the snare's power in this range, and add a broad boost of 3dB or so. Don't be afraid to boost more if it sounds good, just be sure to check the tone on a few different speaker sets to be sure you're not overdoing it.

Good old acoustic guitar strummin' on the low range of the neck is full of 100–250Hz energy. If a guitar sounds "thin" a little boost here can go a long way, especially if balanced by a mild cut in the range where it sounds harsh. Just a few dB on either side can change the tone dramatically.

Organ and key pads with a lot of energy in this range can swallow up other elements of the mix. Vocals recorded very close to the mic may need a little EQ cut from this range.

If you find yourself carving a lot of EQ out of this range, consider arrangement again; by simply inverting a chord part (be it on the guitar, keys, or a wall of vocal harmonies), you can quickly clear up an overly thick sounding mix and improve the arrangement at the same time.

Alternately, if you have an arrangement with many individual parts (keys, guitar, strings, and vox harmonies), consider not having in all the parts all the time! By trading off which part comes in each verse, chorus, and so on you will also keep the listener interested through the variation in tone this provides. A guitar and an organ playing the same riff can be split up in this way; it's a popular arrangement technique, and it's very effective.

Food for Thought: Chord Inversions for Dummies   If you're not sure how *inverting a chord* works, here's how to do it.

Imagine a basic C major chord on a piano. The root position is the notes C, E, and G, in that order. Let's say that string pad is playing that chord while a guitar strums the same chord. The string pad may mask the guitar's strumming. Just playing the string pad an octave (or two) higher may help, but it's not the only way to spread them out.

That low C can simply be kicked up an octave, giving you the chord E, G, and C (up an octave). This is called the *first inversion* of C major.

If you now kick the E note up an octave, you get the *second inversion* of C major.

This works with all chords, major or minor, triads or extended chords (seventh, ninth, thirteenth, and so on), suspended and "add" chords, clusters of notes, and so forth. Additionally, different inversions of the same chord all create their own unique impression, especially extended chords. You may stumble upon that "magic" sound that makes your arrangement shine.

When working with MIDI, you have the endless, non-destructive flexibility to edit the actual notes played in a part, so experiment! Guitarists are able to invert their chords easily by using a capo—so keep one handy in your studio, and learn how to use it.

300Hz–450Hz: Mud

Granted, "mud" is a pretty harsh word to use on a whole range of sound, but it often happens that cutting in this frequency range just makes things clearer.

This is one of the few true "rules of thumb" that you can apply when EQing your tracks. A 3–6dB cut is usually a good idea on guitars (especially acoustic), kick drum and toms, keys and pads (don't overdo it), and low percussion. Use the "boost and sweep" method with a narrow bandwidth to identify the ugliest of the mud and cut away. In this case, use a narrower bandwidth than usual (a Q of 2.5 is a good starting point) because you don't want to cut too much of the neighboring frequencies that give the sound its "body." I find that close mic'ed drums can take a lot of cutting in this area, and it really helps the sound gain clarity and impact. The one exception is snare. Actually, if you're going for a "trashy" snare sound, you may even want to boost in this range. On the other hand, don't cut too much on overheads and room mics in this range, or the drum kit may lose body.

Vocals don't benefit from cutting in this range; they tend to become thin and unnatural sounding. Overall, try not to use heavy EQ on vocals. Stick to broadband adjustments of just a couple of dB, if anything. Our ears are so attuned to the tone of vocals, that EQ quickly becomes obvious. It's best to "shape" the tone through microphone choice. This obviously doesn't apply if you are going for an affected sound. In that case, it's up to your imagination.

Additionally, piano, brass, and woodwinds don't take well to cuts in this range in general. Use your ears; if a cut sounds good, it is good. Strings, on the other hand, can benefit from a healthy cut in the 350Hz area; they start to really shine and become very produced sounding. On the other hand, thin-sounding instruments can benefit from a boost in this range, gaining a sense of weight and girth that gives them more "size" in the mix.

500Hz–700Hz: Body, Warmth, and Chunk

The bodies of many acoustic instruments have resonances in this range that are defined by their physical shape. Acoustic and resonator guitars, dobros, banjos, clavinets, cellos, violas, and violins; think of the unique shape to these sounds.

Because of this, EQ adjustments on these instruments in the 500–700Hz range tend to sound obvious very quickly. A resonator guitar, for instance, is made to resonate in this range—if you find yourself wanting to cut a lot here, maybe you should rethink whether you even want a resonator guitar in the mix! Of course, mellowing out a particularly prominent peak in the range may be just what the doctor ordered.

On the other hand, electric/electronic instruments take well to manipulations in this EQ range. By creating a few narrow boosts in here, a keyboard pad may take on acoustic instrument-like characteristics. Inversely, cutting out of this range makes things sound more synthetic, because musicians tend to associate this range with the "body" of these acoustic instruments.

Electric guitars take well to boosts in here, gaining a certain "chunk" that works well on rhythm parts whether clean or heavily overdriven. For those of you who love a raging distorted guitar sound, this EQ range is your secret weapon. Try this on a double-tracked distorted guitar part: Add a narrow (Q of 2) boost of 3dB at 500Hz to one, and the same boost at 700Hz to the double. Adjust to taste. The complementing boosts on the two parts add up to a much "chunkier" sound than EQing both in the same way.

Lead lines gain "body" and "warmth" from a mild boost in this range more than from boosts in the 100Hz range. Reach for the EQ here if you want that lead part to stand out in the mix—keep it natural sounding by using a wide bandwidth.

Vocals can benefit from a mild, broad boost here if they lack body, but be very careful, or you'll send your singer into the oatmeal-can telephone booth before you know it.

In general, be cautious of heavy EQ in this area on non-electric/electronic instruments; you're entering the range in which human ears are particularly sensitive.

800Hz–3kHz: Critical Midrange

Our ears are most sensitive to frequencies around 1kHz. This is the range wherein much speech-related energy lives, and our ears are made to pick this up. Why do we cup our hands around our mouths if we want to make a point of being heard? Because cupping your hands creates a resonant space, causing a strong boost in this frequency range. Your now-annoying voice cuts right through the soup of surrounding noise.

Load a finished song into your DAW, create a narrow EQ boost, and sweep it around while playing back the song. This is the ouch range. It's not that it's a bad range of EQ that should be cut out, it's just that human ears are very sensitive to this range, and a 3dB boost here sounds subjectively louder than a 3dB boost at, say, 100Hz, or 12kHz.

What does this mean for engineers?

Caution: EQ boosts and cuts in the 800Hz–3kHz range are going to sound much more drastic than adjustments made elsewhere. This also means that a problem resonance in this range in your recording or mixing room should definitely be treated acoustically. Fixing the root of the problem will be a much smoother sounding solution than trying to compensate with EQ.

Interestingly, you can use your sensitivity to sounds in this range to manipulate depth of field in mixes/between instruments. Here are a few tips:

■   I find that if a *reverb patch* draws too much attention to itself, it can be made to "fall back" in the mix with a broad, mild cut in this range on the reverb return. This lets the warm low mids and lows and the "sparkle" through, without competing with the intelligibility of the vocal.

■ A *vocal double* can benefit from a mild cut in this range. It lets the original stand out more while maintaining the thickness of keeping the overall level of the double the same as the main part.

■ *Strident background vocals* can be cooled out a bit with a cut here. Strings behave the same way; even a 1dB cut can make them sit well in the mix.

Again, *caution* is the key word here. You shouldn't be adjusting more than a few decibels unless you're going for a special effect or correcting a serious problem.

3kHz–8kHz: Harshness and Definition

As you have learned by looking at the Shure 57s frequency response, some mics have a built-in boost somewhere in this range. This has the effect of making them stand out in the mix. The problem is, if everyone stands out, aren't we all just shouting?

If you find that a sound is harsh or fatiguing to listen to, your problem lies somewhere in this range. The "boost and sweep" method will lead you to the offending frequency quickly. This is, however, a range you should be very critical of when making your mic selection. If a singer has a "nasal" voice (lots of 2kHz or so), try to pair them with a mic that doesn't bring out this range. Or get a new singer—dang!

In all seriousness, pairing, let's say, a guitar amp with some serious "sizzle" distortion in this range with a Shure SM-57 is bound to create a harsh tone. Trying to correct this with EQ is going to be a tough task, as setting the bandwidth wrong will cut out a lot of neighboring frequencies, leaving the sound dull or hollow. You're much better off keeping your hands off the EQ and in the mic cabinet.

I've made the mistake too many times in the past of working on a mix until my ears were fatigued, making them desensitized to these upper-mid frequencies. I would then start to boost a little here, and a little there. The next day, with recovered ears, everything sounded harsh and the mix was a loss. The moral of the story is twofold—don't work when your ears are tired out, and boosting in this range is just going to make things harsh.

If you find it hard to pick the right mic to keep things from sounding harsh in this range, you may want to try a ribbon mic. Many ribbon mics are ruler-flat in this frequency range, letting you adjust the instrument tone to taste, and then pick up what you hear. This is especially true for guitar amps, strings, and bright-sounding acoustic guitars. If a condenser mic is sounding too harsh on a vocal, try a ribbon mic, too.

Because boosting the high mids can make things harsh, and cutting can rob your sounds of definition, clever engineers have developed a neat trick. By sensitizing a compressor to the frequency band that sounds harsh, you can create a dynamic EQ that only cuts the frequency when it starts to act up. This prevents the loss of clarity that a straight up cut causes, but still ensures against hard peaks. You'll learn how to do this in the next chapter, and even a mild application of this effect can be a life saver.

10kHz–16kHz: Sparkly and Airy, Tinny and Wimpy

Because digital recording media don't exhibit the high frequency loss that analog tape always did, boosting highs has lost a lot of its appeal. Instead, many recordists opt for plug-ins that simulate analog warmth; the exciting effect of a mild distortion that brings out even-order harmonics, making the high frequencies sound both more present and smoother at the same time.

A touch of typical high shelving EQ can benefit a lot of different sources, but keep in mind that, all things being relative, boosting all the highs above a certain point is like cutting all the lows below the same point. This is where adding "sparkle" starts to become tinny or brittle sounding. It's a fine line, and it's always better not crossed. If you don't start boosting the highs on all your tracks, you will maintain a healthy, defined midrange relative to the high end. The mastering engineer will also be able to add the right amount of "air" across the entire mix without any worries about creating a thin, wimpy sound.

Remember that the classic "exciter" effect was created by using EQ and compression on a copy of a track, and blending a small amount of this back in with the original. Here's how you can simulate the classic version in your DAW without an exciter plug-in:

- Create a copy of the track. In this example, let's use a vocal track. Copy any plug-ins as well, especially if you have a pitch-correction plug-in on the vocal, because the pitch and timing might not line up properly if the copy is missing these.

- Next, add a parametric EQ and a compressor to the copy track. Lower the level on the copy track; the initial adjustments may not sound very nice on their own!

- Solo the copy track. On the EQ, boost 10dB at 12kHz with a Q of 1.5. Set the compressor for a fast attack and release, dial in a ratio of 4:1 and lower the threshold until you see 15–20dB of gain reduction. (If this makes no sense to you, no worries; you'll learn more about compressor settings in the next chapter.)

- The copy track will now sound flattened-out and sizzling with high end. Perfect. Lower its volume all the way, and take it out of solo mode.

- Listen to the original vocal again, in the context of the mix. Slowly raise the fader on the copy track with the new effects. As you fade it in, it will sound like the original is becoming more "present" and "airy."

- Compare this effect with simply adding a high shelving EQ to the vocal. Chances are, you'll prefer the exciter effect in a pop or rock context; it creates a certain vibe that regular EQ doesn't. It seems to expand the range of the vocal tone without thinning out the sound as easily as EQ does.

Try playing around with low-pass filters for interesting high-frequency effects. Sweep a low-pass filter all the way up to 20kHz, and move it down the frequency spectrum while listening. Many sounds can benefit from this treatment; there is a certain shape that a filter gives a sound that other EQs can't. Pianos, percussion, guitars, strings, pads, and bass; just about any instrument is a candidate. Again, don't overdo it, but use it to feature particular parts.

Rule of thumb for 10kHz on up: Digital recordings usually don't need much of a boost above 10kHz, especially because many condenser mics have a built-in boost up here. If you're hearing too much from the get-go, switch mics. Remember, ribbon mics provide a smooth, clear high-frequency response, and are particularly effective on vocals, strings, and guitars of all sorts.

16kHz and Up: Attention Canines!

Although human hearing extends up to 20kHz, most microphones don't pick up frequencies all the way up to 20kHz in an even fashion, and most playback systems don't reproduce that high anyway.

But this doesn't mean that you should ignore this band in the EQ. High-end gear designer Rupert Neve is of the opinion that the interaction of frequencies even above the human 20kHz hearing limit may affect lower frequencies through complex interactions.

Although you may not specifically hear a 6dB boost at 20kHz, the filter's interaction with lower frequencies is not to be ignored. This leads to a famous EQ trick, which works particularly well on drum overheads.

Sweep the frequency setting of a parametric EQ all the way up to 20kHz. Start with a narrow bandwidth. Boost by 6–10dB. Now, widen the bandwidth until the filter affects frequencies down into the audio signal's range. This type of EQ boost behaves much differently than a high shelving EQ boost. High shelving boosts all frequencies above the corner frequency by the same amount. By using the bell-shaped curve of the parametric EQ band, you create what is a smooth boost from around 10kHz, which *increases in level as the frequency increases!* So, those frequencies, which generally seem softer at the same amount of gain increase, are boosted more. A perfect high EQ technique is illustrated in Figures 6.11 and 6.12.

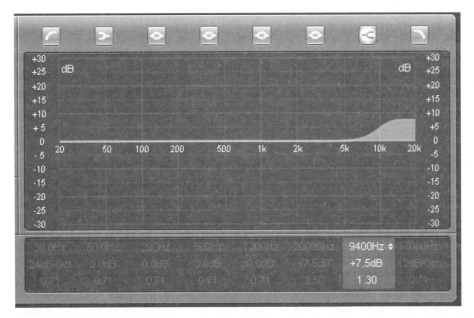

Figure 6.11   Here is a high-frequency boost using a high shelving filter. The downside is that the filter also somewhat boosts "harsh" high mids. It could also stand to give you more of the extreme highs, which are harder to hear at the same amount of boost.

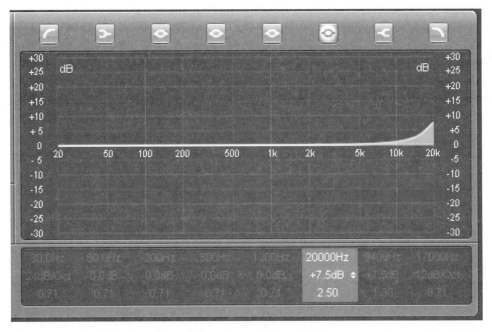

Figure 6.12   By using a parametric band swept up to 20kHz, you create an optimal EQ curve—less boosting of harsh high mids, and more boost of the highest of highs without exaggerated boosting around 10–12kHz, which may sound brittle.

Final Thoughts

No one can tell you on paper how to EQ your tracks. Even audio examples are pretty much nonsense, because it's not *your* audio. Every time you EQ a track you'll use a slightly different setting, and you should use your ears every time. When in doubt, listen, listen, listen—on every set of speakers you can manage.

To help you learn, try exaggerating; boost more and cut more than you think you should, and see how that sounds. Make a rough mix and play it in the car. Compare it to mixes you know. After a while, you'll learn to roughly identify frequencies by ear, and become quicker at making adjustments, as well as know how much to adjust.

In practice, well-recorded tracks (good mic choice and placement, good acoustic environment) won't usually need much EQ, so if in doubt use less EQ rather than more when it's time to make that final mix.

Be particularly skeptical of settings saved by the software designer who created your plug-ins. Rarely does something like the "Power Electric Bass" or the "Pop Mastering EQ" setting on your plug-in actually apply well to your mix. How could it? That setting existed before your song did!

I prefer to create settings based on which filters are activated, such as "2 band parametric" or "1 band + LPF." This lets me grab the filters I want easily and start tweaking the settings right away.

7 Signal Processing Toys: Compression

J ust as equalization allows you to manipulate sound in the frequency domain, *compression* allows you to manipulate the *dynamics* of sound.

As much as the frequency content of a distorted guitar tone differentiates it from the sound of a flute, a piano, or a drum, every sound has particular characteristics that define its audible "shape."

All About Dynamics

The leading edge of a sound is its *attack*. A drum has a strong attack, as do the hammered strings of a piano or plucked strings of a guitar. Bowed strings have an attack that "ramps up" the playing volume, and like woodwinds and brass instruments, this gradual attack can be controlled by the performer. Synth sounds can be programmed to have a very gradual increase in volume, giving the impression of no real attack at all. In any case, keep the term *attack* in mind as the "start of the sound."

Immediately after the initial attack of the sound, the overall level usually decreases somewhat. This is called the *decay* of the sound. Think of a piano striking a note—the percussive impact of the hammer on the string is somewhat louder than the body of the note. The amount that the level of the sound reduces and the time that this takes is the decay.

The body of the sound is its *sustain*. For some instruments, this is also controlled by the performer (on a piano, guitar, woodwinds, and brass), and for others there is a natural sustain that is inherent to the instrument (with bells, drums, xylophone, marimba, harpsichord, and thumb piano). The instruments with a certain inherent sustain can, of course, be controlled by physically damping the sound (by throwing a big, wet towel over the bell after ringing it . . .), but in any case, there's a sustain, even if it's a modified one.

Finally, the sound tapers off into silence. This is the *release* characteristic of the sound. A piano's sustain pedal allows the strings to ring, undamped, thus creating a very long release. Without the sustain pedal depressed, the felt dampeners calm the strings down in a hurry. A guitarist can quickly silence the strings, or allow them to fade out in a long release, and a cellist can continue bowing a note for a long sustain, and when she stops bowing, the strings will vibrate briefly, creating the release of the sound.

Figure 7.1 illustrates these "Attack, Decay, Sustain, Release" (ADSR) portions of sound dynamics. In any DAW, you can see the ADSR contour of a sound by looking at the graphic representation of an audio file.

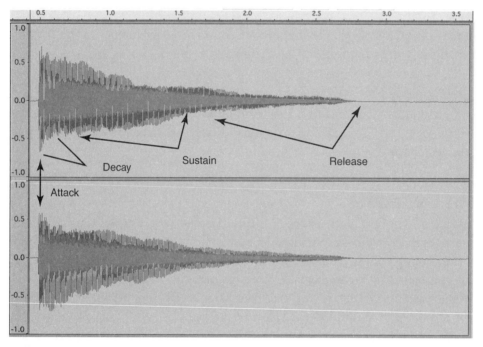

Figure 7.1   By using Audacity to open an audio file of a single piano note being played, you can clearly see the ADSR portions of the sound. The initial attack is fast, and rises to the highest level of the audio file. The sound then decays quickly by a few dB and sustains for a couple of seconds. At this point, the sound releases over another second or so into silence.

Now, let's look at the controls on a typical compressor. See if you notice anything familiar in the naming of the controls, as shown in Figure 7.2.

You may have noticed that two of the four controls are called Attack and Release. Through the application of a compressor, you can adjust the dynamics of a sound, changing its overall dynamic "contour" of ADSR. You should experiment with these controls on some of your tracks to get a feel for how they function. For example, a fast attack on the compressor actually reduces the volume of a sound's attack more quickly (the compressor "kicks in" quickly). Likewise, a fast release allows the gain *reduction* of

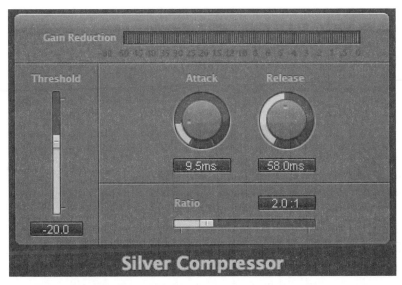

Figure 7.2   The Silver Compressor in Apple's Logic offers top-quality dynamic control. Did you look at the names of the two main controls?

the compressor to *return to less reduction* more quickly, thereby *increasing* the apparent sustain of the sound. It might seem a bit counterintuitive until you get the hang of it. Practice makes prefect!

What a Compressor Does

Let's first learn about the basic controls of a compressor, and how these relate to the dynamics of sound.

A *compressor* does exactly what it sounds like; it *squeezes* sound. It means that it keeps sound from getting louder by automatically turning down the output volume as the input volume increases. It's basically an automatic volume control.

Although some modern hardware compressors and plug-ins may have a few more controls than appear on the simple Silver Compressor shown in Figure 7.2, every compressor has the same basic controls. The additional functions will also be easy to understand if you understand these basic controls of any compressor:

- *Compression Ratio:* By how much does this "automatic volume control" turn down the signal? That's based on the ratio. It's simple. A 2:1 ratio means that for every 2dB over the threshold, only 1dB will come out. *Squeeze!* Ratios vary from 1.1:1 incrementally through the often-used 2:1, 3:1, and 4:1 ratios, to 10:1, even 20:1, and all the way up to Infinity:1, which means that it would take a nuclear detonation to raise the input signal by 1dB. Hope you bought a good mic.

- *Threshold:* Signals below the volume threshold are not affected by the compressor. In this way, you can set it so that only particularly loud signals are turned down. This is helpful in allowing, for instance, a singer to perform quite naturally, only catching particularly loud notes and turning them down. It is also possible to set the threshold very low, meaning that all the audio will be compressed. By then turning up the output volume, the average level of the audio is increased—this is a typical method used to make sounds "loud." You'll learn more about the ups and downs of doing this in Chapter 12, "Mastering: The Final Adjustments."

- *Attack:* Measured in milliseconds, this is the amount of time the compressor takes to start turning down the volume after the threshold is passed. A long attack time can accentuate the natural attack of a sound by allowing more sound through unaffected before turning the volume way down. It can also be set to be very fast, cutting off the attack of a sound, or at least not artificially accentuating it. The decay part of the sound is also affected in this way; after the set "attack time," the compressor kicks in, quickly reducing the signal's level, thus artificially creating a decay. I'll explain this with some real-world examples in just a second to clear it all up.

- *Release:* After the compressor kicks in and turns down the signal, and when the input signal is no longer over the threshold, the compressor is allowed to stop turning down the audio. This is usually not done instantly, as there would be an odd "rushing" sound as leftover sound suddenly becomes louder. If, however, you do choose to set the release time short, you can often make the sound seem to have more sustain. The lower-level "sustain" and "release" part of the audio will *increase* in volume as the gain reduction *decreases.*

- *Peak vs. RMS Mode:* Some compressors allow you to choose between Peak and RMS (root mean square—simply put, a mathematical method of averaging values) operating modes. In Peak mode, the compressor responds to fast transients more quickly, whereas RMS mode looks at the average signal level, tending to ignore short peaks. These modes respond somewhat differently than the general attack and release settings, and you should listen carefully to determine what works best on your track.

- *Knee:* This control determines how the compression ratio behaves as the input signal crosses the threshold. *Soft-knee* compression is characteristic of many classic compression units, especially some of the old favorites that used optical elements to control the compression. Soft-knee compression provides a smooth transition as the compressor begins working, and is consequently a great choice for vocal compression. *Hard-knee* compression means that the compressor abruptly begins working at the set ratio as soon as the signal crosses the threshold. A hard-knee compressor's action is therefore more audible than a soft-knee compressor at a given ratio. Many compressor plug-ins allow you to select between hard- and soft-knee settings. Let your ears be your guide!

Just to keep things straight, compressors don't actually turn up signals, although they can make them seem louder. The output control of a compressor can increase the overall gain of the signal, but the action of the compressor is always that of turning the signal down. What makes things seem louder is the increase of the average level of the sound. This, along with the ADSR properties of sound is important to understand.

Think of it this way: A guitar plucks a note, and that initial attack is rather loud, but this quickly decays into the much lower level sustain and release of the sound. The average level of this sound over time was not much affected by the short attack and decay phases of the sound. See Figure 7.3.

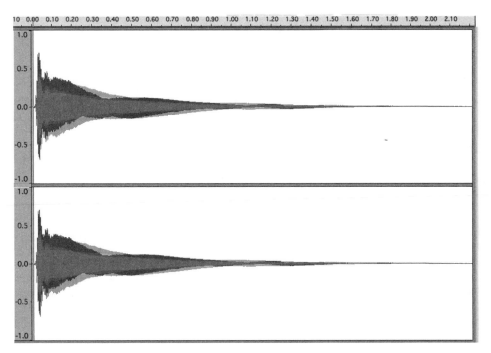

Figure 7.3   Here's the uncompressed guitar note. Notice the defined ADSR elements. Let's fatten this little note up.

Now, play that same note through a compressor. By setting the threshold low, you will "dig down" into the sustain portion of the sound, first allowing the compressor to release as the sustain fades into release. At this point, the compressor releases its volume control, returning to unity gain. However, you've also cranked the output! So, the overall level is increased, and it sounds like the note has more sustain—it's louder for longer! See Figure 7.4.

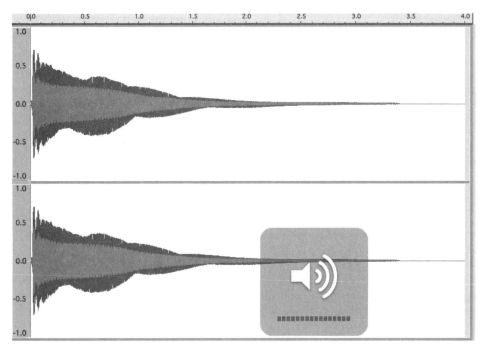

Figure 7.4 Here's the little guitar note after a Snicker's bar sundae diet with whipped cream on top. By decreasing the attack, setting a fast release, and cranking up the output gain, the average level is now much louder, and the sound sustains a higher level for a bit longer. This is a rather exaggerated example, but it illustrates what compression can do to your sounds.

INSIDE THE BOXES

A Basic History of Compression   Originally, compressors were designed to take care of the technical problem of audio levels getting too loud for the recording medium and ruining the recording. So a band rocking away without a compressor may go all the way up to an ear-poking, tape-overloading level, but with a compressor set in there, musicians can set a limit to how loud they want to let things get.

In the old days, musicians needed to record things at a high level to avoid losing the signal in vinyl crackling and tape hiss, but not so loud as to cause the stylus to skip or the tape to distort. It also happened that compression just seemed to make things appear louder—it was a way to squeeze those loud sounds onto the limited dynamic range of vinyl or tape.

The FCC also has strict standards set for radio station output, and peaks higher than a station's registered output can result in stiff fines. Naturally, radio stations also need to be loud to be competitive, so they try to push the signal level as high as possible without going over. In this case, limiters set the bar. Modern FM stations use multi-band compression to compress different frequencies bands of the audio spectrum separately, pushing average levels higher and higher. You may have wondered how some stations can get so loud. There you go.

Current recording media has plenty of dynamic range available, especially when using 24-bit audio. We could do away with a lot of this "level chasing," since we seem to have the extra dynamic range to fit the music in anyway. It just happens that the tone of compression, as well as its "loud-ifying" effect still appeal to listeners. Certain classic compressors shape the tone of instruments or mix in pleasing ways. The Neve 2254, Teletronix La-2a, and Fairchild 660 are world-class compressors that have shaped so many well-known songs, that the sound of these units often brings that "big studio" edge to the mixes they compress. Luckily for users, these often *very* expensive units are now available as affordable plug-ins!

Limiting and compression also make music competitively loud (that is "as loud as the other guys on the radio"), for better or for worse.

Limiting

The meaner, bigger, tattooed brother of the compressor is the limiter. In fact, a limiter is just a compressor with a really high ratio. Modern limiter plug-ins also often have the ability to simply tell the plug-in, "Don't go above this level. Ever." Dutifully, the hard-knuckled fist of Big Brother Limiter will pound down the head of any signal before it can cross the level you set.

Limiters are powerful tools for keeping your tracks from going over *digital zero*, the maximum volume in the digital domain, and the point at which nasty clicks, pops, farts, buzzes, and other artifacts can occur.

This may seem like an invitation to slap a limiter across all your tracks, and not have to worry about levels any more. Unfortunately, this thinking has become all too common, and audio quality can suffer tremendously because of the careless use of limiting.

You see, the way a limiter works is to flatten out that signal before it passes the level you determine and creates a digital "over." The problem is, this flattening of the signal also makes it pretty harsh to listen to! Anything more than a few dB of limiting starts to make your tracks or mix sound hard or harsh. This sustained, high average level also makes for a fatiguing listening experience. In many cases, the listener will just want to turn the music down. So, in the end, your quest for more volume through limiting and compression ends in what many engineers are now calling the "wimpy loud sound;" a high average level mix with no natural dynamics. It sounds the same whether played loud or soft. It's missing one of the basic requirements of music—dynamics. For more info on this, please go to http://en.wikipedia.org/wiki/Loudness_war.

Finding a balance between loudness and dynamics is a very important consideration for you as an engineer. The dynamics of your music contain as much of the emotional impact as the lyrics, melody, and harmony. When mixing and mastering, you need to decide when it's "loud enough" without ruining the dynamics through the heavy-handed application of just one effect.

Not to take the wind out of limiting, not by any means! It is a powerful, useful effect that, when applied correctly, does give your music an edge.

Food for Thought: Understanding Levels and Meters   Compression and limiting give you a unique control over the level of your signal, allowing you to control both short peaks and the overall volume of your audio tracks. It is important to understand how audio is measured in order to properly record and process your music.

If you are a DAW user, the meter you most often see is probably the dBFS meter. The "FS" in dBFS stands for full scale and refers to the maximum numerical value that can be digitally encoded. If you are recording at 24-bit resolution, a 0 on the dBFS meter corresponds to 24 bits of all ones—the largest number that can be represented by binary numbers in the 24-bit digital format. Think of the reading on the meter as being a feather floating on the ocean; it doesn't tell you much about the volume of water contained in the waves, but rather shows you their peaks and dips very quickly.

However, this doesn't mean that the level of the incoming audio signal will not exceed the voltage that corresponds to 0dBFS! This is why it is important to understand levels, how they relate, and what your meters are actually telling you.

In professional audio gear, an analog signal of approximately 1.23 volts (RMS) is calibrated to correspond to about –20dBFS. This voltage is also referred to as the level +4dBu—a marking you have probably seen when looking at the connections on your gear. To keep this in perspective, when you adjust the gain knobs on your mixer or mic preamplifier, this affects the voltage, and therefore the level on both the dBu and dBFS scales.

However common as it may be in DAWs, dBFS is just one way of measuring the level of your signal. A third scale also comes into play, and it refers to the old analog standard for measuring audio levels; the VU (volume units) meter. That's the meter with the needle on it. The way that the needle reacts to the audio levels provides a sort of averaging effect to the metering. Returning to the ocean analogy, imagine a log floating in the water—short wave peaks don't have the "oomph" to push the more massive log around. The log bobs along, riding higher when waves with more volume push it up. This is how the VU meter needle responds to audio. (This does simplify things a bit—VU does have its quirks at different frequencies.)

What is important, though, is that the VU meter better represents what people *hear*—the average level over short periods of time. You just can't cross that 0dBFS line for short periods of time, or else . . . ZZZZT!!!

There is no standard for converting between analog and digital levels; it varies from country to country and sometimes from gear to gear. In the US, −20dBFS is specified as 0VU, whereas in Europe, −18dBFS is specified as 0VU. Many audio converters allow you to specify your own calibration (my Pro Tools rig uses −18dBFS as an alignment level), so be sure to check the default settings on your gear.

The trend in recent years has been to calibrate the meters differently (again, they are not fixed to one another) so that 0VU sits higher on the digital scale, at perhaps −14dBFS or even higher. This increased average level leaves less headroom before the ZZZT! happens. Consequently, engineers who opt for a higher average level (more "loudness") are forced to limit the audio more severely. This in itself can lead to a more fatiguing, harsh-sounding recording.

It is good practice to use both dBFS and VU metering, and to calibrate your system for both. As a home or project studio engineer, you should also reference recordings similar to the style you are producing, comparing levels with those as a sort of watermark to aim for. Copy a song from a CD into your DAW, play it back, and get to know the way the levels behave. How far are the peaks (dBFS) from the average (dBVU) levels? Do you think that the song sounds compressed or limited? If so, do you find this to be *musically* fitting, not just technically, from an engineer's standpoint? Considering these aspects will keep you and your ears fit to produce music for years to come.

Let's imagine a recording of a drum kit, loaded with percussive sounds that create very short signal peaks that may approach (or even want to cross) 0dBFS. Snare drums, for instance, can spank out some high peaks, even when set with a conservative recording level. These short peaks don't represent the average level of the drum kit, which includes all the tone of the drums, not just those short peaks.

Although 24-bit digital recording provides plenty of dynamic range for cutting your basic tracks without your audio falling down into the noise floor of your gear, compression and limiting can help you get a handle on levels as well as shaping instrument tones in a pleasing way.

Here are some tips for proper limiter use:

- By "shaving off" some of the short, fast peaks in the audio that are all but inaudible, you can then raise the average level, making the whole shebang louder. So if you use a limiter to reduce the peaks by 3dBFS, you can then raise the level of those drums up 3dB on the digital scale. This doesn't mean you should raise the level—keep in mind, you may be adding a veritable orchestra of other instruments to the track, and their cumulative level will probably push the whole mix level up so far that you'll be turning those drums back down. Drums, percussion, acoustic guitar, acoustic piano, and brass instruments typically have short dynamic peaks that can be safely reduced by 3 or 4dB without getting into the "loudness overload" range. Just don't overdo it or you'll flatten the performance.

- Sometimes seriously mangling audio with the "crush" of a limiter is the effect you're looking for, and the tried-and-true example is drum room mics. The effect of a high compression ratio, a low threshold, and LOADS of output gain makes it sound like those room mics are sucking up all the tone they can get out of the room.

- Get a feel for how using a limiter to reduce fast peaks can reduce the difference between peak (dBFS meter) and average (VU meter) level. See how far you can go before you hear the sound start to get "hairy" (where sounds approach some type of distortion). Find the point where things get "hairy," and then back off a bit. Compare this to the unprocessed audio. Listen and learn...

 Again, check out Chapter 7 on the CD-ROM for some "before" and "after" audio samples of room mics, and how to set that up yourself. Keep clear of those mics when they're cranked up; they're libel to suck the hair right off the top of your head.

Expanders and Gates

The trusty compressor, who turns the audio down when it goes above a set level, is still in heavy use in studio music production. I dare say more than ever. However, there are a few relatives who also had their places in the studio years ago (and may still show up in your plug-ins list) but who no longer hold a very important position in the studio toy chest.

The *expander* does the opposite of what the compressor does in terms of volume ratio. When a sound goes above a certain level, the expander turns it up. Now, what good could this possibly do?

It turns out that expanding was an integral part of noise reduction in analog tape days. A compressor was used on the way to tape to boost the overall level above the noise floor. On playback, the opposite ratio was used (at the same threshold) to then de-compress (expand) the signal back to its original dynamic range. Result: Tape hiss lost and dynamic range maintained. This is still useful if you use analog tape, but otherwise an expander may only be useful if you happened to over-compress a track, and want to undo the damage. If you know the original compressor settings, just dial in the opposite ratio (that is, 1:8 instead of 8:1) at the threshold you had used and hope for the best!

Gates (and *downward expanders*) are the opposite of compressors in terms of threshold. A compressor turns down the audio when it goes above the threshold; gates turn it down. This means you make quiet sounds quieter. If you have a lot of background noise, you can use a gate to cut the signal off when there is no musical signal, removing the noise from your tracks. A gate cuts the noise out, a downward expander reduces it, again, by a ratio making the action somewhat smoother to listen to.

The problem with gates is that this abrupt "cutting off" can actually draw attention to itself, defeating the purpose of keeping things quiet. In this respect, a downward expander is better (they are often the same unit, just that a gate has a very high ratio, like a limiter, whereas the downward expander has a milder ratio).

A typical application for gating is on tom tracks, where there is constant bleed from the kick, snare, and cymbals, with tom fills showing up only once in a while. However, now that just about every DAW has (at least) mix volume automation, You can do a much more accurate job of cleaning up those tom tracks with volume automation, or by simply cutting (editing) away parts of the file where there are no tom hits. Just be sure to leave space to fade in and fade out around the tom hits, so that the edits themselves aren't audible.

Compression Tips, Tricks, and Techniques

Compression uses can vary from the subtle balancing of relative instrument volumes, to in-your-face effects, as you just heard with the room mics. In some cases, the effect is simply a tool for proper recording technique, in other cases it is a splash of paint in itself. Let's check out a few examples, learn a few settings, and listen to what you're able to do.

Drum Compression

Modern pop and rock drum sounds almost always require some compression to make them "larger than life." You just need to be cautious not to squeeze the very life out of them! Drums need a strong dynamic element, or else they will sound flat and artificial. Here are a few tricks for fattening up the sound without ruining the natural dynamic of your tracks (refer to Chapter 7 on the CD-ROM for audio examples):

- *Dig deep, but keep the ratio low:* To "thicken up" a drum sound, it is possible to compress everything, not just the dynamic peaks. This works to increase the average level of the drums, making them sound thicker—it brings out the body and tone of the kit. The trade is a loss of dynamics, so this is not a set formula, but a subjective technique to be altered to fit your song.

- *Trim the peaks:* As mentioned earlier, by reducing the fastest transients just a bit, you can gain more level for the track without adversely affecting the dynamics. Combining this with a mild "dig deep/low ratio" compression can create a louder drum sound that maintains punch and still sounds subjectively loud.

- *Classic drum compression:* Back in the old days of one- or two-mic setups for drum recording, the drum sound was put entirely through one compressor, simultaneously working on kick, snare, cymbals, and so on. In doing this, the attack and release times of the compressor were critical adjustments. If the compressor kicks in and out too fast, it can pump the signal. *Pumping* is when the action of the compressor is audible in a bad way. By adjusting the attack and release time carefully, the compressor action can be heard, thereby subtly controlling the loudest hits, pulling up the body of the kit, and still letting enough dynamic through to keep the kit sounding lively.

■ *Individual pieces:* When you have kick, snare, toms, and so on, all on separate tracks, you can shape the dynamic of each piece individually. By taming some of the extreme peaks on individual pieces (especially the kick and snare), you can also ensure a smoother compressor action of the buss compressor for the whole kit—one renegade hit won't "pinch" the whole kit.

Tip: This tip goes for any individual piece of the kit and the whole drum submix, and has been credited to Frank Zappa, although I'm sure many other engineers in his era were onto it. *You can compress the same tracks two different ways!* Zappa used to send the drums to two different busses, and label them "fat" and "crack." On the "fat" buss, he would use the "dig deep, low ratio" technique at a more extreme setting, which brought out the "fat" body of the drums. On the "crack" buss, he boosted the highs and used a compressor with a slower attack, accentuating the dynamics. When mixing the song, the overall drum tone can be balanced between "fat-body" and "dynamic-crack" easily. Just beware of the "fat butt-crack" sound.

To set this up, just assign your drums to two separate busses, and set the compressors and EQ as described. It's also possible to use aux sends to do this trick if your DAW doesn't have buss assignments. Just use a pre-fade aux send, and insert the compressor on that aux return. This concept is similar to the Motown-style exciter setup.

This technique can help you achieve a huge snare sound, for example. First, duplicate the snare track (or use a pre-fade aux send with another compressor inserted). On one track, adjust the compressor for a "dig deep/low ratio" setting until you hear loads of body and fatness. On the copy, set a slow attack, medium release, and a ratio of 3:1 or so. Adjust this until you hear tons of "crack," and not as much body. Now you can adjust the relative levels of these tracks in your mix to get the optimal snare sound.

Instrument Compression

Compression settings, as with EQ, must be adjusted to the song, individual taste, and to the performance of the part. Consequently, there are no hard and fast rules. Here are some rules of thumb for different instruments, and some things to watch out for.

Factory settings and plug-in presets tend to be lousy, as they are meant to be very audible in music-store demo rooms, and cannot—obviously—be optimized for any track you have, as they cannot hear your track and make an informed decision.

The best way to learn is to experiment, listen on many different monitors, and most importantly, compare your work to professional recordings. I have found that song intros that feature just acoustic guitars, just drums, and so on, make excellent ways to compare your mix to the pros. You may want to sample a few parts of songs that you like and keep them handy for comparison. I cannot provide these on the CD-ROM, for copyright reasons, but you should be able to put together some favorites of yours in an easily accessed folder. Use a CPU-friendly audio program like Audacity or Real Player to open these examples in the background while working on your DAW; this makes for easy comparison without disturbing your session.

Some tips for different instruments:

- *Percussion:* I have found that shakers, tambourine, maracas, and the like, sound best either not compressed or crushed to death. (Note: Congas and larger bodied percussion can be treated as drums when it comes to compression.) A well-performed shaker part will sit in the mix with no problems. Due to the fast transients of percussion, compression doesn't really do much other than turn down the volume of each shake. This is why the other, cool-sounding option is to crank the compression to the hilt. This brings out loads of room tone as the transients are pulled down and released quickly, "bringing up" the ambience. Give it a try—use a fast attack and release, low threshold, and a 8:1–10:1 ratio.

- *Acoustic guitars:* With their strong transient leading edge, you have to be careful not to use too fast of an attack on acoustic guitars. I've made that mistake, and they become impossible to mix; the lack of attack ruins their rhythmic element, but they still seem loud since the body of the sound remains. If you want a natural sound, be sure to leave some attack in there! Start with a 4:1 ratio, slow attack, and medium release. Listen to the guitar in context. If it sounds like it needs more body, speed up the attack and increase the output gain. You'll find that listening to acoustic guitar along with drums will help you find the right amount of attack. If you cut off too much of the dynamic, the guitar won't blend well with the drums—they should dance together.

- *Piano:* If you do record acoustic piano, you'll have the option of moving the mic closer to the hammers for more attack, farther away for less attack and more sustain. If you use compression with a very fast attack and release on piano, you can seriously cut down the attack of the hammers, creating an almost synth-like sound. This can be a brilliant effect or a bad mistake, depending on the context.

- *Overdriven electric guitar:* A certain amount of compression belongs to the sound of electric guitar. Although an overdriven amp naturally compresses the sound, an outboard compressor can tweak the tone for more attack, body, or sustain. Heavily overdriven sounds are already pretty compressed by nature, but can give you even more sustain. By using a compressor to pull the level down when a note is played, the volume is "dammed up" by the compressor, and as the note dies away, the compressor releases, bringing the volume back up and giving the sound more sustain. When you do this, don't cut the attack too fast, or you'll lose definition. Crunchy guitar parts benefit from classic compressor settings (3:1, med-slow attack and release, dig the threshold in for 4–6dB of reduction). Tweak attack and release times while listening in context, as with acoustic guitars.

■ *Clean electric guitars:* These guitars love compression as well. Sustain can be earned the same as with overdriven guitars; dig deep to dam up the volume, and set the release to give you volume as the note fades. You'll find that attack settings are especially easy to hear on strummed chords—clean electric guitars are great to experiment on when learning the ropes of compression setting.

■ *Bass:* Before reaching for the compressor, check out the bass in the mix. Many good bassists don't need much compression at all; they have a feel for how much attack and sustain they need. Too much compression on a bass guitar can reduce the bass to a big rumbling mess, and lost attack is hard to get back. With that in mind, start with a 4:1 ratio, and back off if you seem to be losing attack. Likewise, keep the attack slow to allow the definition of each note to pop through.

Vocal Compression

Vocals require particular care, and I won't claim to be an expert at vocal compression. It's a fine line between "not enough" and "too much," and I'm pretty much never satisfied with vocal sounds until I get away from the mix for a few weeks! Perspective definitely helps...

It is very easy to over-compress a vocal, which is the worst thing that can happen. An over-compressed vocal seems never to sit in the mix; always detached and floating above the music. Some engineers swear by manual level control—keeping a hand on the vocal fader and riding it through the song. Naturally, this can be automated.

Riding the fader, if you think about it, is like a compressor with a moderately slow attack. As the engineer hears the vocal level increase beyond what they hear as an acceptable level, they pull the fader down a bit. When the level drops, they push the fader back up. It's intelligent compression! They'll also be prepared to allow an overall higher level in, say, the chorus than the verse, although they'll still pull down loud peaks. This can be likened to adjusting the output of the compressor.

So with this concept in mind, it's best to err on the side of slow attack and release when setting vocal compression. Take advantage of plug-in automation, and automate parameters if you need to. For example, raise the compressor output in the chorus, or reduce the ratio in spots where the compressor clamps down too hard on dynamics.

A few more tips:

- Basic vocal compression. Finding the right amount of compression for a vocal track can be difficult for new engineers. Here is an easy way to find a good compression setting. Start by listening to the vocal in the mix, with no compression. Set the level so that the verses sound correct. When the chorus approaches, the singer is probably going to get louder, and this will make you want to turn down the vocal a bit, so that it doesn't overwhelm the mix. Determine at which point you feel the vocal needs to be turned down, and use this to set your compression level. Start with a 2:1 ratio. Play back the track from the verse, and as the chorus enters and the vocal gets louder, reduce the threshold so that the compressor starts working at this point. Rewind and listen again. Adjust your threshold and ratio to find the right amount of compressor action to make the vocal "sit" in the track.

- The "dig deep/low ratio" concept works great on vocals, increasing the overall sense of presence as subtle sibilant sounds and breathing are brought up. This may be a problem if you have a sibilant/pop-happy singer. In this case you may want to try the duplicated track idea here as well. On the first track, set the compressor to dig deep, with a fast attack and release to bring up body and closeness, adding a de-esser to control sibilance/pops. On the copy track, use a higher threshold, 2:1 or 3:1 ratio, slow attack and medium release to allow the natural dynamic through. Dip down the worst of the sibilance with volume automation so the vocal doesn't sound heavily processed the whole time. You'll need to use your ears to be sure things don't start sounding unnatural.

- You can also use two compressors, one after another, to deal with different aspects of the vocal dynamic. The first compressor is set to a "classic" setting to generally make the vocal more present (low ratio, low threshold, and slow attack and release), and a second compressor is set to catch drastic volume changes. The second compressor should have a higher threshold (again, "kicking in when things get really loud"), with a faster attack and higher ratio. This two-tiered compression scheme can sound more natural if you need to tame an inexperienced singer.

- Although slower attack and release times sound most natural, if you want a really in-your-face sound, go extreme. Dig deep and use a high ratio, cranking the output. The singer will climb out of your monitors and grab the listener by the throat. Just keep in mind that these extreme "loud" effects can make people want to turn the recording down, defeating the purpose of making it loud.

■ Keep the "Motown effect" in mind—it was originally invented for vocals! A quick reminder: Duplicate the track, compress the original mildly, squash the duplicate, and add a high EQ boost. Turn the duplicate down, slowly add just enough to add an "exciting" effect to the main track.

Room/ambient microphones are top candidates for compression. By compressing a distant mic, you bring up the lower level reverberant sounds by reducing the peaks of the direct sound that reached the mic, and then turning up the output. This is, of course, key to drum room mics, but is also useful on ambient mics used on vocals and any instruments. Keep the attack fast to prevent the compressor from accentuating the direct sound, which can sound "slappy" in recordings made in small spaces.

 Because listening is more important than reading in this case, be sure to spend some time listening to the examples in Chapter 7 on the CD-ROM and experimenting with compression and limiting on your own.

Luckily, most DAWs allow you to use plug-ins on your tracks without affecting the recorded audio permanently. If you find yourself running out of plug-in power, be very cautious about using destructive (written permanently to the audio file) compressor and limiter processing; they are very difficult to undo! Instead, always save a copy of the basic track without the processing (perhaps on another track or playlist) before using destructive processing. In this way, if you overdo it, you will be able to return to your original uncompressed audio.

Now go have some fun squeezing, crunching, and smashing tracks into compression heaven!

8 Signal Processing Toys: Reverb- and Delay-Based Effects

Now you come to what is, for most engineers, the really fun part of effects processing. EQ affects the frequency domain, and compression/limiting deals with dynamics. In that same vein, reverb- and delay-based effects work in the *time* domain.

Whether it's an extremely short time, as with phasing and flanging (a few milliseconds), or long, interstellar delays and luxurious reverbs (progressing over seconds), these effects are created by time-delaying a copy (or many copies) of the original audio signal. Pitch can also be manipulated in the digital realm, allowing for harmonizing, chorusing, and other pitch-modulation effects. Some delays can also be set to play back the delayed samples in reverse, opening the doors for a variety of effect combinations.

Sometimes EQ and compression are used in combination with delays to create the desired result, as is often the case with reverb. The original tape-based delay effects created in the 50s and 60s had a natural filtering effect, caused by analog tape. Years later, engineers are still trying to simulate that sound with digital delays, EQ, and compression!

Keep these two points in mind when dialing in effects:

- *Always use your ears to judge effects, and go with your gut reaction.* If a track sounds great with no plate reverb, let it be. Run a rough mix and listen. If a delay sounds better when not locked to the song tempo, go with it. Impression of the sound takes precedence, not what the controls show.

- *Be skeptical of presets.* As usual, settings in any outboard processor or plug-in are meant to impress potential customers, not to serve the song. Not to mention that a setting called "Indie Rock Reverb" is probably the last preset you should use when mixing an indie rock track. How independent/individual can it be if it is preset?

This chapter starts by breaking down each effect into its basic elements. From there, you'll understand what constitutes a reverb, a chorus, and an echo, and you'll be able to create effects that serve your tracks perfectly. At that point, save your own presets—you can alter them from song to song, but they'll become part of your unique production style. A little knowledge in this area can make the difference between a boring track with typical, stock effects and a track with an edge of its own.

INSIDE THE BOXES

Auxiliary Sends and Returns   Although using *auxiliary sends* (aux sends, for short) is more a topic for Chapter 11, "Mixing: Balancing Art and Craft," it's best to clear up the concept now, as it will be mentioned often when discussing delay and reverb effects.

An aux send is simply an additional way to output a duplicate of the signal from a track. Generally, your track's audio is being sent to the main mix bus (stereo output), which you are then feeding to your monitors. It's also possible that your tracks pass through a sub-mixing bus, which is then routed to the main bus, as often happens with drums. This allows you to control the entire group's volume with just one stereo fader.

Let's say you want to add reverb to several tracks, creating a general sense of space for the drums, bass, percussion, and rhythm instruments.

It would take an awful lot of reverb units (or plug-in instances) to put a reverb on every track to which you want to add reverb!

A better way to do this is to create one reverb instance (or use one outboard unit) for all the tracks that need this effect. Then, you send a copy of the signal to that reverb unit, and return the reverb-only signal to the mix.

This is where the aux send comes in. On your mixer, you connect aux 1 to the input of the reverb, and the reverb's output to either a dedicated aux return or a stereo channel input—either of which is then sent to the main mix bus.

The reverb's output should be set to "100% wet," meaning that the output is only the reverb signal, and none of the original input signal. If you were to insert a reverb directly into a track, instead of on an aux send, you would need to adjust the reverb's wet/dry mix so that you still hear some of the original signal, and not just a wash of reverb. Figure 8.1 shows the setup for a basic reverb send.

By setting up effects on aux sends, one effects unit or plug-in can process a collection of tracks, saving you processing power, and keeping the overall level of the effect output easily manageable. This is the most typical way reverb is used. Phase, chorus, and flange effects often sound best when inserted directly on a track, instead of on an aux send. However, by placing the effect on an aux send, you can process the effected signal again, separately. You should experiment to discover what works best on your own tracks.

Caution: Should you choose to use channel inputs for the reverb return, be careful not to turn up the aux 1 send on the return channels! This would send the reverb's output back to the input; at a certain point, this feedback loop goes crazy and—ZEEEEE!— you'll get ear-splitting feedback. Used with caution, however, a touch of reverb feedback can create an effect with a different sonic texture than just increasing reverb decay. Play around, but at LOW volumes!

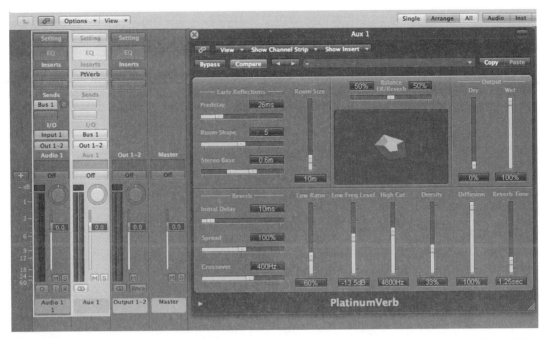

Figure 8.1   The aux output on the audio track is set to send to bus 1 and the aux input channel has its input set to bus 1. By raising the level of the aux send on the audio track, a copy of the signal is sent through the reverb plug-in on the aux input.

Delay: The Building Block of All Effects

When you go to add a delay-based effect to a track, the goal is to make the sound fuller, more spacious, insanely "scream-a-licious," or something along those lines. In any case, the sound will wind up bigger than the original. In layman's terms, this is done by repeating the original signal, often many times, and doing something to those repeats.

A basic delay effect has a few basic parameters, and these parameters show up in various incarnations on just about every delay-based effect, so it's good to understand just what they do:

- *Delay Time:* Simply the amount of time between the original sound and the delayed signal. This is usually measured in milliseconds and can vary from a few milliseconds up to several seconds. Longer delays generally require more processing power. Keep this in mind when creating delay plug-in instances on host-based processing systems (such as Pro Tools LE). Use the plug-in with the shortest maximum delay time that suits your needs. Longer delays with the delay time turned down still eat processing power!

- *Mix/Level:* Depending on how you apply the delay, you'll want to adjust accordingly. If you create a delay on an aux send, set the "mix" to 100% effect, and adjust your level via the aux return. Tip: Always be sure that you are sending enough signal level to the effects units/plug-ins. As with any audio device, they sound best within a certain operating range. Check that the input peaks around –6dBFS, and adjust the level you want to hear via the output, not the input. Low input signals lead to noisy effects outputs or, in the case of reverb, ineffective operation—the full reverb effect may not completely develop if the input is too low.

- *Repeats/Feedback:* To create multiple repeats of a sound, a bit of the output of the delay is fed back into the input. The more you put back in, the more you get out— and on and on... With very short delay times, this can turn into a shrill acoustic feedback-like effect, so beware. Chorus and flange effects are intensified by turning up the feedback, but may start to take on a particular "hollow" sound if too many extremely short delays stack up and cause frequency cancellations.

- *Tempo/Note Value:* Some delays offer the option to synchronize the delay time to note values within the song. You may be able to designate the tempo within the delay's interface, or, as with many current DAWs, the tempo parameter is linked to the session's song tempo. This makes dialing in 1/4 note, 1/8th note, 1/8th note triplet, and so on, delays very simple.

- *Filters:* Although you can easily run your delay output into an EQ plug-in, many delays have built in high- and low-pass filters. It's usually desirable to filter some of the lows and highs from the delayed signal, as full-fidelity repeats of your track may sound confusing in the mix—digital delays can be perfectly crisp and clear, making them quite distracting! The classic echo effect is best achieved by filtering lows and highs and keeping the level of the repeats below the level of the track itself.

- *Modulation:* An interesting thing happens when you change the delay time while a signal is passing through the delay unit—the pitch changes! This is the principle behind chorus effects. A mild modulation of the delay time (longer-shorter-longer-shorter) occurs, and the pitch of the delayed signal varies slightly. This can also give life to echoes, especially on guitar and synth solos, but be cautious—it draws attention to itself. To create a chorus effect from a delay, set the modulation and then reduce the delay time to a just a few milliseconds. Set this way, it seems as if the source signal is modulating because you can't hear a distinct echo. See Figure 8.2.

Flanging effects are created by blending a pitch-modulated signal with the original signal. If you have a delay that can be set to 0 ms, set the blend to 50% and listen to the flanging effect occur as you adjust the modulation.

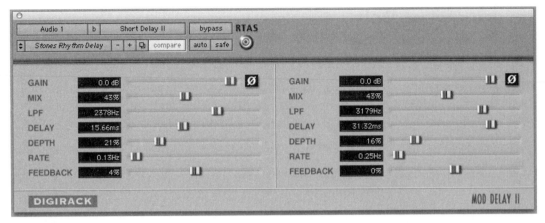

Figure 8.2   Every DAW has at least a basic delay plug-in available. Pictured here is the Pro Tools delay plug-in, an amazingly flexible tool that allows users to create a surprising variety of delay, panning, chorus, and flanging effects. Pictured are basic settings for a rock n' roll rhythm guitar short stereo delay. Because it is inserted directly on a track, and not an aux send, the "mix" is not set to 100%.

Food for Thought: Subtle Delay Time Adjustments   Before there were "smart" plug-ins linked to your song tempo, delay times had to be calculated by doing a bit of math. This is still important if you're using an outboard delay unit, and for making minor adjustments to the "feel" of a track.

Calculate a basic delay setting by dividing the number 60,000 by your song tempo.

For example, a song at 120BPM, that will give you the magic number 500.

The number 60,000 comes from 60 seconds in a minute × 1000 milliseconds in a second = 60,000 milliseconds per minute. When you divide by beats per minute, you get the delay time in milliseconds per beat. (Remember high school math? Cancel the units!) Divide that by 2, 3, 4, and so on, to get useable short delay settings, and adjust by ear for "feel."

Important trick: A delay time that's a bit shorter than the beat division "pushes" the track. The delays coming in ahead of the beat give the track a nervous energy. Times that are a bit longer sound more "laid back," always falling in behind the beat. This is an important aspect in working on the feel of a track. Minor adjustments can make a huge difference in this case!

As noted in Chapter 1, "Looking Back for New Inspiration," Les Paul pioneered flanging by synchronizing two tape machines with copies of the same tape. During playback, he'd hold the side of his hand against one reel, and flanging was born. The name "flanging" refers to the edge ("flange") of the tape reel!

Reverb: Delay, Delay, Delay, Delay, Delay

Artificially created reverb comes in a few different forms. This section focuses on digital reverb, although it's important to know how the other types function.

You already read about the "reverb chamber" concept earlier in the book; simply feed a track back through speakers into a large room and record the natural *room reverb* sound. All physical reverb processors are variations on this idea.

Spring reverb feeds the audio signal into a transducer attached to metal springs, and picks the signal up with a second transducer at the other end. This is the typical guitar amp reverb. *Plate reverb* uses the same concept, but the transducers are attached to a large, thin metal plate suspended in an (typically very large and heavy) enclosure. *Hall reverb* is what you'll hear in a concert hall with good acoustics. It's pretty unlikely that anyone at all has an empty concert hall lying around that they can use as a reverb chamber. I've recorded choirs in unfortunately dry-sounding spaces, and made the clients quite happy by adding a touch of "concert hall" reverb from a Lexicon PCM-90.

The clever home recordist can take advantage of spaces, amps, and instruments to create some groovy natural reverb to spice up recordings. A piano with a weight on the sustain pedal becomes an amazing reverb tank! Just open the piano, place a small monitor at one side, and a mic at the other. Play your track into the piano, and record the results. You can play with EQ on the signal sent to the speaker to get certain strings ringing. You can lay a T-shirt on the strings to dampen the reverb if the decay is too long.

Tiled bathrooms are great ambience generators, and are worth mic'ing in stereo for the best effect. Long hallways and basements also work well, although they may suffer from a rather "boxy" timbre. A guitar amp with the reverb cranked to 11 makes a great natural spring reverb generator. Send an aux out to an output on your audio interface, and plug that into the amp input. Mic the amp, and return that signal to a track. Record the reverb signal to that track, and you have your reverb—no need to set up the "aux send to amp" every time you listen!

Digital Reverb

Digital reverb is simply a simulation of various spaces and classic studio devices, created with many, many delays. The delays are so densely packed that you don't hear them as individual repeats, but as a "wash" of reverb.

By manipulating the amount, level, spacing, and EQ of these delays, different simulations of spaces/devices are created. Generally, there are some basic settings that get you into the ballpark of how "room," "hall," "plate," and so on, should sound, and you adjust from there.

INSIDE THE BOXES

Convolution Reverb   One of the great recent developments in reverb effects is convolution reverb (sometimes referred to as *impulse reverb*). Audio Ease's Altiverb, Waves' IR-1, and DigiDesign's TL Space are three popular plug-ins—TL Space is relatively inexpensive. The basic way the reverb sound is created is different than with standard digital reverbs, which use multiple decaying delays. Convolution reverbs use a complex mathematical formula (convolution algorithm) to simulate physical spaces.

A convolution reverb must have an *impulse response file* loaded in order to function. An impulse response file is an audio recording of a space, made by recording a short burst of white noise—like a starter-pistol shot—or a sine wave signal swept through the audio spectrum (the impulse). The sound stimulates the acoustics of the recording space (that's the impulse response) and the recording captures the results for the plug-in to use for processing.

There's no need to get into the math behind convolution reverb; it suffices to say that you can sample any acoustic space. Guitar in a barn? Let 'er rip. Piano in a penthouse apartment? Tickle them ivories. Vocal down the toilet? Bombs away.

There are many audio enthusiasts out there who've recorded collections of impulses for use with convolution reverb. Just Google the name of your convolution reverb plug-in along with "impulses" or "impulse response" and you'll have more impulse response samples than you can ever use. There are great ones and lousy ones among the freebies, so you've got another rainy day project—sorting impulse response samples. You can find impulse responses from plate reverb units, classic outboard reverb units, concert halls, rooms, and even some famous architectural spaces.

Additionally, convolution reverb can be used to sample speaker cabinets and other resonant objects. This is a key aspect in amp simulator plug-ins, and means that you can design your own unique virtual amp by combining plug-ins (for example, EQ, compression, distortion, and convolution reverb). You could also use a sample of the resonance on an acoustic bass or piano to make a synth bass or piano sound more realistic.

It turns out that you can load any audio file you like into the plug-in. The output may not sound much like reverb when loading something that's not an impulse, but you never know what's going to happen, so try it out. A piano chord as an impulse for a drum part? How about the other way around?

New sounds abound—don't think of the convolution reverb as just a processor for creating ambience—it can also be a speaker simulator, filter, vocoder, or turn-my-sound-into-a-science-project processor.

As with delay, reverb has a typical bank of parameters that you should become familiar with. Here are the most important tweaks to make when searching for optimal ambience, and how they affect the sound:

- *Pre-delay:* This is the difference in time between the original sound and the onset of the reverb, usually measured in milliseconds. There is always some sort of pre-delay in natural spaces. In a small room, it may be just a few milliseconds, whereas in a large concert hall it could be in the hundreds of milliseconds. If you find that a reverb sound wells up too quickly and crowds the source sound, try lengthening the pre-delay. You may even want to make the reverb enter an eighth note value after the initial sound, for a more rhythmic, musical effect.

- *Early reflection:* This portion of the reverb provides the spatial cues our ears re-cognize as the size and shape of the space. In any space, the sound source reflects off the walls, floor, and ceiling, and makes a quick return to our ears. These reflections are typically very short echoes, and their level and tone quality gives us a sense of the type of material in the room as well. A tiled bathroom has very loud, bright early reflections. In a concert hall, these reflections will take a little longer to arrive (due to the size of the space) and will likely be diffused by the design of the room to prevent distinct echoes from interfering with the timbre of instruments. A carpeted living room will have early reflections with less high-frequency energy due to the absorbing qualities of carpets, curtains, upholstered furniture, and so on. Plate and spring re-verbs, due to their artificial nature, lack early reflections, making them suitable for creating sustain without a sense of room.

- *Decay:* Decay is the length of time it takes for the reverb sound to trail off. This is often referred to as RT60—the time it takes the reverb level to decay by 60dB. It is important to carefully adjust reverb decay—too short and the reverb is ineffective, too long and the sound becomes muddy and indistinct. The trend in recent years has been toward shorter reverb times. In the 80s, many recordings were loaded with long reverb tails (especially on drums and vocals!) for dramatic effect. Current production trend is to create a sense of "space" and "room" without long artificial sounding reverb trails. Some engineers claim that reverb only muddies up a mix and leave it out entirely, preferring to work with delay and natural spaces for ambience.

- *High-frequency (HF) damping:* In a natural space, high frequencies are absorbed more rapidly than lower frequencies. This aspect is dependent upon the materials in the reverb space. A stainless steel tank will bounce high frequencies around a lot longer than a carpeted living room. Try adjusting this parameter to control the "brightness" of the reverb and, consequently, its "attention getting" factor. Sounds with more high-frequency content simply grab attention more. If you are happy with the length of a reverb (decay time) but it still sounds too obvious, try increasing the HF damping. If your reverb sounds "muddy" or indistinct, try reducing the HF damping for a brighter, clearer sound. You may then need to lower the level or the decay of the reverb to compensate.

- *Reverb density:* The multiple late reflections (after the louder early reflections) that make up the decay of the reverb are, naturally, separated in time. In a "denser" reverb, these reflections occur closer together, creating a smoother reverb well suited to drums and percussion. In a lower density reverb, the reflections are more distinct. This will often sound pleasant on vocals and non-percussive instruments, but may sound "choppy" or "grainy" on percussive sounds.

- *Diffusion:* This parameter relates to the behavior of the decay portion of the reverb. In layman's terms, the more diffuse the reverb, the smoother it sounds. The disadvantage of high diffusion can be a loss of the sense of "room" and "shape" in the reverb. A room reverb typically has a low diffusion, a plate or concert hall reverb a rather high diffusion. If your reverb gets cloudy and indistinct sounding, again, reduce this parameter to gain shape and clarity.

Figure 8.3 shows Logic's PlatinumVerb.

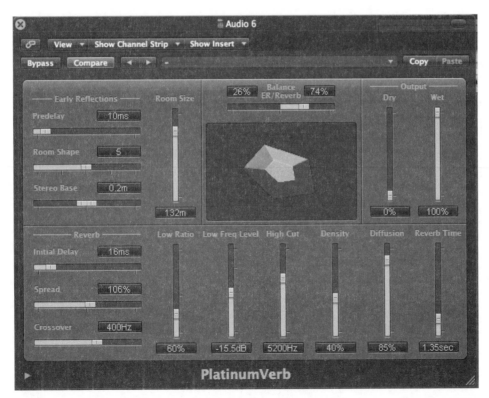

Figure 8.3   Logic's utility reverb PlatinumVerb offers easy adjustment of every parameter, and includes a visual representation of the room shape the reverb simulates. Be sure to adjust the output to 100% wet if you use the reverb on an auxiliary send.

Now that you know the elements of delay and reverb effects, it's up to you to come up with interesting, functional combinations.

Consider the possibilities:

- A rockabilly guitar track with a plain eighth-note digital delay vs. a delay that's "pushing" the beat, and filtered through an amp plug-in with a twang of spring reverb.

- A vocal track slathered with standard plate reverb and echo versus a vocal with a tight, present ambience, followed by a brief wash of modulated reverb that enters a "lazy" quarter note later. The reverb's output could be modulated with chorus, flanging, pitch shifting, or any other effect in your arsenal.

- Snare drum with a simple room reverb vs. the same snare fed into two aux sends— one sent into a very short delay to create stereo spread and then sent through a short spring reverb, the second compressed heavily and sent into a "tape emulator" plug-in pushed into overdrive. Your basic snare track will become a monster.

- A single background vocal with plain plate reverb vs. the same reverb split left and right, pitch-shifted down a fifth on the left and up a third on the right.

- Simply placing a delay before a reverb input allows you to manipulate what the re-verb hears, allowing you to adjust the stereo spread, pre-delay, and even modulate the in-going signal in ways that the standard reverb plug-in may not allow. Daisy-chaining plug-ins often leads to unusual and inspiring results.

 I've provided examples on Chapter 8 of the CD-ROM for you to listen to. Not every track needs a fireworks display of crazed effects to make it stand out. Always consider just leaving a track dry, or using a combination of direct and ambient microphones to create a natural-sounding space.

9 Perspectives on Music Production

There is no simple definition of music producer. Traditionally, the producer was responsible for organizing the recording budget, booking studios, communicating with the record label, coaching the musicians, and overseeing the development of the songs from rehearsal to mixing and mastering. The modern producer may wear many suits—musician, engineer, financial overseer, manager, mixer, or other jack-of-all-trades.

Some producers take a very heavy-handed approach, adding their mark to all things creative; others steer the ship according to prevalent winds, providing the musicians much leeway in the studio. The best producers know that there is no definite formula for a successful production, but they always manage to bring structure and confidence to the project. They can also deal with time, budget, and technical constraints in a way that keeps the music in the driver's seat.

If you want to successfully produce music, you'll first need a way to deal with the technical side in a way that doesn't distract from the creative side. This axiom applies to group efforts and to do-it-yourself productions.

Getting Ready to Produce Music

Let's first consider a project involving a group, which is more complex than if you are producing tracks by yourself on a DAW and sequencer. Most important is delegating studio time according to budget. There is nothing less inspiring than running out of studio time when tracking vocals, overdubbing, or even worse—having to rush through a mix, and being unhappy with the mix. Even if you want to produce a project on spec in your own studio, you need to be clear as to how much time you are willing to spend on the project, be it per day, per week, or for the entire project.

Here's another checklist for you to consider. Naturally, the studio setup part (from Chapter 4, "Setting Up Your Studio") comes first. If you've already set up your studio, or if you are recording elsewhere, you're ready for the next set of tasks. If you can get these issues cleared up at the outset, you'll be free to concentrate on creativity and quality:

- *Songs:* Are the songs in shape? Perhaps they will develop during the recording process, but there should be a strong lyric, melody, and harmony to guide you. Don't rush into the studio expecting to create there. It can and has been done, but if you are constrained by a budget for studio time, this could be a serious problem.

- *Performers:* Has the band rehearsed? Are the performers ready to work in a studio setting? Even seasoned stage performers may become flustered quickly if they have had little studio experience. Consider this when choosing a studio (or setting one up). The monitoring situation is critical. If the performers will be using headphones, they may require separate mixes, which not all studios can handle. Do your homework! It is also possible (á la many Rolling Stones recordings) to set up the studio as if it were a club, with a PA, monitors, and all. This may be a better method for cutting basic tracks. Discuss this and determine how the musicians will be most comfortable.

- *The budget:* Money, as politics, can make for strange bedfellows. Be clear and realistic about the budget up front, and speak about it with the band. Consider splitting the budgeted studio time into parts: separate basic tracking, overdubbing, mixing, and mastering budgets. Vocal recording and editing can be particularly time-consuming, so factor that in.

Consider using different studios for basic tracking and for overdubbing. A larger, more expensive studio may offer the necessary amenities for tracking drums and other basics in a more efficient (and better sounding) way than a small studio that does its best to cram you all in. The increase in quality and time saved may keep you on budget, despite a higher hourly rate.

This applies to the home studio in equal measure—consider your time versus the quality you may gain by tracking for even one day in a better equipped studio. Instead of buying a drum kit's worth of mics, invest the cost of just two of those mics in getting some killer basic tracks. After that, you can overdub yourself silly with just a couple of mics.

Discuss your budget with the studio owners/managers; they may be accommodating if you have a clear budget plan, and offer you some more time for a "lockout" rate. That is, if you're planning on spending 12 hours tracking basic tracks, they may be ready to give you two full, eight-hour days at the same price.

Does the budget also need to pay for CD printing and advertising? How about your fees as a producer?

- *Additional gear:* In a way, this falls under budget, but you may be better off purchasing some things on your own rather than having a studio provide them. For example, ask the studio which hard drives are acceptable. Find out if the studio provides any amps, guitars, keyboards, organs, and/or if they charge for this—and budget gear loading costs if you have to bring your own. Strings, picks, drum heads, drum sticks, and other "perishables." Analog tape, if you are going to use tape. Brainstorm for more before the recording dates. Same goes for the at-home recordings; hit the stores before you start the sessions so you don't repeatedly interrupt the flow.

- *Personnel:* Some studios require you to hire a house engineer and assistant (this is typical in and around Nashville, LA, and NYC). Some studios may allow you to provide your own. Consider hiring someone who knows the room! The time and effort saved, at least when cutting basic tracks, will probably far outweigh the few dollars you may save by doing the engineering yourself. If you are hiring session musicians, be clear on their fees, and find out if you must pay double scale for a bandleader!

- *Extraneous details:* You all need to eat, sleep, and get to the studio. Even if you're close to the studio, budget in the use of the bassist's van, and get him some gas money regularly; this will prevent hard feelings (he could be in the band *because* he has the van). Don't let any particular financial burden fall on a band member or other member of the production uncompensated—they may not say anything for a while, but trust me, it will come out in one fight or another. Best to get it all cleared up before the problem starts.

All in all, the preparation side is not too hard. A couple checklists, a folder for receipts, a few phone calls, and a visit to the studio, and you'll have a clear plan. Now you can get to the part you've been looking forward to—being locked in two windowless, sound-proof rooms with a gang of artist for several weeks. Wait. Help... I wanna go home!

Finding Your Place as a Producer

So, how do you start producing? The uninitiated might think it's all recording and mixing tricks, with some sweeeet arrangement ideas, and suddenly they're tellin' every-one what to do and rich and famous hangin' in a Hummvee limo?

Okay, so why not? If that's your goal, give it a shot. No one's stopping you. If you can find artists willing to back up your big-shot ego trip fantasies, more power to you. But you're more likely to get a prompt introduction to the door.

The truth is, every producer has a specific style, so there is no formula for success. Many producers simply grew up with the star musicians with whom they work, and grew into the job by taking care of the many details specific to those musicians!

This also means that those "So you wanna be a music producer?" ads all over the web are just a slick way to get money from fools. If you want to be a successful producer, the word is "Action."

Get organized, get serious, and be ready to spend a lot of time in the trenches. The public persona of most famous modern producers is a calculated show, a way to sell their image. Behind the scenes, it's serious business. Management, composing, rehearsing, performing, recording, mixing, advertising, promotion, touring, video shoots, publicity, and so on. Producers get involved in various stages of creating the music, depending on their abilities.

You have to find your own strong suit, and stick with it. You *must* be a good communicator; a producer is not a star, but a coordinator, liaison, motivator, and creative problem solver. Some producers don't even touch a mixing console, although they know what they want to hear. Some become part of the band, playing guitar, or keys, or co-writing and arranging songs.

So what do you do? Read on to find out.

Becoming a Do-It-Yourself Producer

You may be preparing to start a do-it-yourself project, in your own home studio. You'll be wearing all the hats—writer/arranger/performer/producer. Although you have the luxuries of your own space, no time pressure, no arguing with egomaniac singers, all your toys at hand, and the creative freedom to write and arrange as you please, here are some points to keep in mind:

- *Freedom* can soon become "freedom, horrible freedom" if you begin drowning in the sea of possibilities offered by the amenities of the modern home studio: DAWs, sequencers, loop libraries, endless plug-ins, and synth patches. The studio becomes solitary confinement. It's always possible to adjust the arrangement, do another take of that vocal or guitar part, and remix the song. Projects tend to never find an end. The best advice I can offer here is to set a time when something must be ready. Obviously this won't work if you only tell yourself, as you could just put it off. So, tell someone that you'll be playing the new songs at the party on "x" day. Whether that's your birthday, a picnic on the 4th of July, Boxing Day, who cares—it's up to you. Just be sure to tell a lot of people, so there's no backing out. The pressure will make you work, and force you to accept whatever mix you have at that time. It's easy to obsess about details, and a set date forces you to finally call it done.

- *Get input.* Invite people in to listen now and then, and ask for opinions. You should hope for criticism! If your partner thinks it's "nice," get another opinion. "Nice" is the little sister of "crappy." You need an honest critique from someone whose opinion you respect. It's unlikely that your work is a masterpiece, although it may be very good. If it's that good, make it a masterpiece—get advice.

- *Manipulate technology* to re-inspire yourself, and have some plans ready to keep the creativity flowing:

  If you're stuck while working on lyrics, try some typical songwriter strategies, like changing the point of view, or removing all the Is, mes, and mines (or adding them if there aren't any). Try snipping phrases out of newspapers and throwing them all in a paper bag—then pull a random line when stuck. Try out the program Liptikl from www.intermorphic.com; a program that blends together sources and gives you new output. I've also had luck just putting text into Google translator, and then translating to Russian, Russian to French, French to Chinese, and then back to English. You never know what that might do.

  It can be helpful to go low-fi to give you a new perspective. Record a demo of the song into a Dictaphone (or other crappy cassette deck). Play just acoustic guitar and sing and stomp your foot. Play this back and listen as if you've never heard it before. All the static and distortion may interact to create a new sound that you can turn into a part.

- *Get out of the boxes.* Sure, it sounds cliché, "think outside the box," but everything is a box! Your computer and the windows of the DAWs and sequencers, the monitor speakers, the control room, the studio, and the house. Take this literally if you run into any blocks.

  When a track is sounding too pure, get the audio you're working on *out of the computer*—run it through an analog mixer or analog effects, re-amp and re-mic it, even record it onto a crusty old tape deck and play it back into the computer. Sometimes DAWs are just too clean and innocent; audio files can gain character when given a beating in the real world.

  Get the music out of the monitors and the studio. Play it on a keyboard in the living room, or on a guitar at singer-songwriter night. Playing the music outside the studio can tell you a lot about what's working and what isn't. Especially consider tempo, key, and arrangement.

  Get your butt out of the house. Too much time staring at a monitor skews anyone's perspective. Take a break for an hour or a whole day. Meet a friend for lunch, go for a hike in the woods, or wander the streets in your city. Clear the cobwebs.

Food for Thought: A Producer Case Study: Mahdyar Aghajani   Rock n' roll has caused plenty of stress between youth and society. Politicians and other finger-pointers have blamed rock, rap, and heavy metal for everything from teen suicides to drug abuse and violence. Court cases, mass record burnings, and attempts to brand and censor albums have occurred at many points within the history of recorded music.

Now, hip-hop—with its fusion of bass-heavy beats, angst-filled lyrics, frenetic collages of syncopated percussion, and dissonant harmonies—is stirring up trouble all over the world. Iran's premier hip-hop composer from the capital city of Tehran, Mahdyar Aghajani, has recently found himself at the epicenter of another shakeup in the world of music and politics.

Composed and recorded in a private Tehran studio under the cover of night, and away from prying ears, Mahdyar and vocalist Hich Kas (translated as "No One"), created the first professional-grade hip-hop album in Iran. The anti-government lyrics of the album, "Asphalt Jungle," struck a note with Tehran's disillusioned youth, making the breakthrough album an instant hit. The first 5,000 CD pressing was sold out within a week. Government officials quickly sought out the artists behind the subversive work, arresting Hich Kas, and forcing the then 17-year-old Mahdyar into hiding. He continued his work while on the run.

As a classically trained musician from the age of six, Mahdyar learned to read, write, and compose music on traditional Persian instruments. By integrating these sounds into his productions, he not only added an unusual depth to his hip-hop tracks, but also gained the respect of the conservative music community. A published interview wherein other composers in Iran condoned Mahdyar's work put the government hot on his trail. He fled to Europe in the summer of 2009 in the wake of rioting in the streets of Tehran.

 I met with Mahdyar in Paris this winter to catch up with him and check out his current work. He continues composing and recording despite the chaotic conditions that accompany living as a political refugee, working with samples, loops, and MIDI instruments to complete his productions. He works on a Mac laptop using Reason and Pro Tools, and has been kind enough to share some of his thoughts about producing music in a portable home studio environment. In Chapter 9 on the CD-ROM, you can hear samples of his work and get an inside look at the cutting edge of international hip-hop.

Mahdyar's recent work includes music for the film "No One Knows About Persian Cats," a docu-drama about the life of underground musician's in Tehran. The film has won several awards, including a special jury prize at the Cannes Film Festival. He continues producing tracks for Iranian and British artists in London and Tehran.

Musicians Producing Music

Many of you will probably be working with others, sharing performance and production tasks. Are you a musician, preparing to produce music? In that case:

- *Be sure you can communicate with both technically proficient and technically ignorant players.* There are killer musicians out there who cannot tell you where middle C is on their instrument, and those who can't invent a part but can play Paganini's 24 Caprises while standing on one foot. Learn to read notes and play a keyboard, at least enough to find all the basic chords in a key and some typical substitutions and outside chords/notes. For the non-technical players, be ready and open to speak their language. Some non-technical drummers, for example, do really well with just a block diagram of the arrangement. Get animated and communicate!

- *Absorb stylistic clichés, and learn to mutate them.* "There's nothing new under the sun," "Art is about art," "Creativity is about hiding your sources," "A musician is a clever thief"—there are endless quotes justifying the recycling of musical ideas. Be prepared to kick out some ideas typical to the style of music you're working in, and to modify them to create something new. Arrange a background vocal in the shape of a funk horn part, play a typical reggae bassline through a fuzz pedal, adapt a typical guitar line for a synth part. Don't fear clichés, just be aware that they have been used, and may need to be altered to sound fresh.

- *Create a library of sounds and make them accessible.* This goes for loop and sample experts, keyboardists, guitarists, or anyone.

If you specialize in loops and samples, have your workstation or laptop ready to go at all times. There's nothing worse than having to start up, scroll through menus, connect cables, and finally hear something. Inspiration can be fleeting—be ready to create music. Programs like Ableton's Live, Propellerheads' Reason, and even Apple's GarageBand are quick and easy ways to work with loops and samples whether making demos or cutting final tracks. Being able to change keys and tempos easily is a big plus.

Keyboard players, sort your patches. Create a folder/library with your most useful basic sounds, and your favorite custom patches. Anyone who has even messed around with a keyboard/synth knows that 90% of the patches are "music store patches" meant to turn heads in the music store, but with little use in real production. As a starting point, I suggest making a copy of all the patches you have, and then simply deleting all the crap. Who needs 17 chorused, cross-delayed variations on a Wurlitzer sample? Pick the cherries and dump the rest.

Guitarists, get out your useful gear, and make sure it works. New 9v batteries, working patch cables, a couple of cool, small amps; these are the things that keep a session rolling. If you have an amp simulator plug-in you like to use, see the keyboard suggestion, and create a small, effective library of useable sounds.

- *Put your ego on the shelf.* Support the song. Keep the focus on the vocal, if that's the point of the song. Less is often more; open space in music is as important as the notes played. Be critical of what you play in equal measure to your critique of others—ask the other musicians for feedback on what you play.

Guitarists, unless you're producing a guitar solo–driven album for a niche market, be prepared to shelve those 32nd note arpeggio sweeps at the first sign that the band doesn't dig 'em. Don't underestimate the power of using an interesting chord inversion instead of the usual open-position or power-chords. Although a simple, crunchy power chord is sometimes just the right thing.

Keyboardists, what you add may often be the glue that holds a track together! To protect this, avoid monitoring too loud so that you don't make everyone doubt that the part fits! In many pop and rock tracks, there are layers of synths "below" the drums, bass, guitar, and vocals helping the track tremendously, but when turned up, they drown out other tracks. I am a fan of these parts, but I have experienced too often that they get voted against when too loud. Singers always seem most nervous about this for some reason.

Additionally, consider your "layering" power—doubling a bass part with a piano can sound incredible, backing a guitarist with the same chords in the same inversions can be just the right thing, and the track may need nothing else. And fight those egotistical six-stringers if they think your rhythm part doesn't fit with theirs. Tell them they're right; you could mute the guitar—whatever serves the song!

Loop geniuses, pile up the loops! Then mute them. Don't place everything you have throughout the whole song. The repetitive nature of loops requires them to be either constant and hypnotic, or surprising and refreshing. If you have one "backbone" loop, let it run on and on—it will fade into the background of the listener's perception. Then by dropping it out, you create a new dynamic—you win back attention. On the flip side of this, using a great loop for just a short time, and cutting it out before the listeners get their fill, is a surefire way to make listeners want to hear a song again.

Keeping the Producer's Position in Perspective

If you're not a musician looking to produce, then perhaps you're more interested in wearing the engineer/producer hat. In this case, also keep your position in perspective:

- *Engineering is technical.* The musicians may not be interested in knowing why you picked a certain microphone for the floor tom, so spare them the blow-by-blow. I've seen too many embarrassing situations in which the audio geek just won't shut up, and it's a drag on the session. It's rare that the engineer is a star, and it's a lousy profession for those seeking recognition. Commit yourself to tracking great sounds that will knock everyone out when they play the mix in the car, or crank it up in the apartment. The sound will impress people, and keep them coming back for more. When you've pulled this off without a peep, people will start asking you how the hell you did it. Then you can have some real fun messing with them.

- *Keep it organized.* This is for your own good as well as the sanity of all involved. Use patchbays so you're not around behind the racks half the time. After setting up mics and cables for a session, use Gaffer tape to secure cables to the floor, so your mic stands aren't pulled down by Bigfoot the Bassist. Have a backup hard drive ready, even if the band has its own. Keep an extra backup—every so often a disaster does happen, and you will be elevated to God-Hero status if you've kept a backup. Also, take session notes; players, amps, instruments, mics, and so on. If someone has to punch in a few notes later, it's good to know who played what on which amp and instrument. A digital camera can take care of this in a few snaps.

- *Focus on the monitoring situation.* You should have your own chops together, and be ready to get great sounds easily. This way, you can spend time setting up the monitor mix in the phones. Nothing is more important from the performer's perspective, and great monitoring makes sessions enjoyable. This, again, translates into a good reputation and repeat business.

- *Listen with the artists.* Once you're set up and can record a test track, listen back and check out the sounds. Solo each instrument (or group) and ask what they think. Be prepared to make adjustments, but keep the initial test take as a watermark. Sometimes you had it right the first time, and by comparing, you can see if the sounds are improving. There's nothing better than a quick comparison to gain an objective view. The drummer may be most difficult to please, so be sure you have a few drum mic'ing tricks up your sleeve. Remember that sometimes simpler is better; if the drummer is not happy with what he's hearing when you play back the overheads and kick drum (the classic drum sound), something else is wrong. Make sure the drums themselves sound good—try different mics, a different setup, or maybe a new drummer.

- *Be the communication center.* This is another way you'll gain respect and more work contracts. Be relaxed and clear about what's happening in the studio; give clear cues as to when the tape is rolling. Make sure even those with their headphones off are up-to-date with what's going on when they come back from the bathroom. Let people know how many bars they have before the punch-in, and please leave a little post-roll! It's unnerving to have the punch-out followed by immediate silence.

Yeah, the engineer works hard at a thankless job. It seems that everyone else gets praise for doing something the way it's supposed to be done, and the engineer only ever hears if something goes wrong. However, when musicians talk about the quality of a studio, they often talk about how capable the engineers are. Many musicians are lost as far as the technical side goes; they need a capable engineer, although they hate to admit it. Do a solid, no bull job and they'll seek you out again and again.

Studio/Session Priorities

The sessions begin. Now your goal as producer is to capture great performances in the studio, at the best quality you can.

With your financial/organizational tasks well under control, and with confidence in what you can bring to the project, you can now focus on the music and recording of it. This is no time to forget the absolute priorities, listed here from most important to least important:

- A well-written and well-arranged song

- An inspired performance

- Good tone from the performer and instrument

- Proper studio acoustics

- A solid monitoring system

- Choice of mic and fitting mic technique

- The recording gear

In recent interviews, venerable producer Rick Rubin, when asked about his working methods, said that he prefers not to work in the studio—that the work needs to be done before the recording begins. He stated the need for the band to work out the songs in rehearsal, to write, arrange, and play them together; that this is the energy that needs to come into the studio to make a successful record. These points are valuable.

Always, I repeat, always, set these priorities straight, in the order shown. Make a cue card and glue it on the wall. Think about a few situations, and you will understand the importance of this order. I suggest them as a guide to all steps of the recording process.

Situation 1

The singer belted out a killer vocal during the basic tracking of a song. The take captures the feeling you all want, but it was tracked with a handheld mic, and has some cymbal bleed and the occasional popped "P" and handling noise. It's time to overdub vocals, you have a $2,000 mic set up, and the tone is great. The problem is, the singer is just not hitting the "vibe." What to do?

The list says that performance takes precedence. The mic choice is subordinate, as is tone. You can certainly try re-recording it, but don't waste too much time there. It is better to have a heartfelt take on the song than a polished turd. A good engineer can probably reduce the background noise and popped Ps anyway. Just come back once in a while and take a stab at the vocal, and see if the singer nails it again—make her a rough mix with the great vocal take and sing along to it; it may bring her back to that vibe.

Situation 2

When trying to overdub a guitar solo, everyone is listening from the control room, and no one is happy with the tone, not even the drummer. You've had the house engineer break out vintage tube mics and class-A preamps, and that should sound great, I mean, what's all that expensive gear for!?

The list says to listen to the amp in the studio. Make sure the source sounds good. Swap guitars, amps, pedals, anything. Get the guitarist in there and tweak it until it sounds great, then record it. Tone from the instrument has precedent over mic and gear. A well-recorded lousy sound is still lousy.

So, you get the gist. It is very unlikely that the order of importance will change. Don't throw out an inspired performance just because of the tone. Don't believe that expensive gear will make a lame song sound great. Always try to improve, but do this by *raising the bar.*

Opinions, Criticism, and Space Madness

Although you should already have a good idea of what the band is capable of, and where there limitations are, you will need to decide when to push a performer to strive, and when they're tapped out. If you've developed a rapport with the player, it's much easier to be critical. A rapport doesn't have to be a friendship; mutual respect is enough. This means being a practiced listener. You are there to *listen* after all, aren't you?

Try to avoid opinion. Instead take a critical approach. Don't misunderstand criticism as a negative thing, but rather as an evaluation. As a producer, you should be searching for the rather elusive aspect of "quality" in every aspect of the music. If a vocalist is sounding overly melodramatic, you need to guide her into a more honest performance. Have her listen with you, and ask her what they think. Ask her to criticize, as she'll need to live with the performance! Evaluate the track together, and don't be afraid to look the vocalist straight in the face and tell her you think the performance sounds melodramatic; but be prepared to offer an alternative.

In any case, the common goal in the studio is good music, despite any small arguments that may arise. Get to know the dynamics of the band. Sometimes one person is the main songwriter and arranger, sometimes all members play a role. Some musicians just act the part of hired gun; they are there to play their parts, and are otherwise rather unattached.

If there are heated arguments, act as a mediator and resist taking sides. If someone gets angry, allow time for tempers to cool off, and then discuss the issues only in the presence of all band members. Should some members leave the room to cool off, they shouldn't feel that there was any talking behind their backs. You can honestly reassure any alienated artist that they were not the target of a character attack in their absence. This gains the respect of everyone.

The studio is a strange animal. Although a song may have been performed many times, it is a daunting feeling to know that something is being recorded for posterity. It can also be unnerving to hear one performance over and over—there is a tendency to hyper-analyze; to find fault where there is none. Both the producer and the performers need to keep those "perfect imperfections" in mind, and decide when they are satisfied with a take.

It is entirely possible that the first takes will be the best ones; they may be the most spontaneous and loose sounding. Be sure to keep them! Minor sins (sloppy drum fill, missed chord change) can be punched in or overdubbed, whereas a basic feel may be hard to get back to. When the analysis of the studio creeps in, the smooth, "recorded" sound may lack the raw feel of a performance, and it is precisely this aspect that gives some artists their edge. Know the artist well enough to decide if you should be trying to capture this, or if more precision is needed/wanted.

Be prepared to take criticism as a producer, as well. You are there to serve the music, and may have to adjust your methods as the project develops.

During a long project of mine, I was reminded of an episode of *Ren & Stimpy*. It is an apt description of what happens when you're enclosed in the studio too long, working on the same songs over and over and over. In the episode, they were in a spaceship, in the grip of "Space Madness;" speaking gibberish, eating soap, approaching mutiny and murder. Needless to say, we needed a break, and were able to get a laugh out of deciding to stop for a bit, rather than opting for the mutiny and the murder. My point? When the unwelcome studio hallucinations begin, it's time to get some fresh air. Stop listening to your tracks for a while. Go see another band perform, go for a hike, visit a museum—recharge your batteries.

Musical Scaffolding

When producing a track, there is often a flood of ideas at the beginning stages—little background vocal melodies, some guitar and keyboard parts that you're not sure will fit, and lots of percussion parts and little counter-rhythms that everyone hears in their heads.

Some producers approach this flood of ideas by telling the band to determine what works by rehearsing the song. The idea is that the cream will rise to the top, and the less useful parts will fall away. But the opposite approach can also be very useful—keeping every idea and experimenting with combinations by mixing.

Sometimes having a big, complex ocean of sound to sing or play to makes the over-dubbing process easier. I think of it as "musical scaffolding;" it may be removed (as desired) at mixing time, but it provides a basis to which you can work on the song while developing the track. This is very useful for developing musical ideas and instrument parts that would otherwise sound out of place if the song arrangement were stripped down to the basic drums, bass, rhythm instruments, and a melody. Ambient sounds, humming and whistling, swirls of percussion and accents, extra rhythm loops, counter-melodies, traffic sounds—anything goes.

Naturally, this requires a lot of tracks, but most DAWs now give you at least 48 tracks (many even more) and MIDI tracks take up very little space and CPU power. So, if the stripped-down version of the song is still missing that edge that makes you really love it, start taking in all the ideas you can put down, and build a scaffolding around the song from which you can climb about and build up a skyscraper of sound.

When you've added all that you can come up with, you may want to mute all but the basics, once again. Make sure the performances are solid. Should the vocalist be striving to rise above the collage of sound, or sitting comfortably within the chaos? Are there parts that obviously collide in the wrong way, or are there "happy accidents" that have created unique harmonies? Now is the time to make confident decisions about these things, and to feature them. When mixing, consider grouping elements that come to-gether well. If you would like to hear this sort of concept at work, I suggest listening to Brian Eno and David Byrne's "My Life in the Bush of Ghosts" and "Everything that Happens Will Happen Today," David Bowie's "Outside," or "Wah Wah" by British indie-heroes James. These are fascinating collages of sound with much to offer to the attentive producer.

The All-Important Rhythm Track

One of the cornerstones of production technique—especially so in hip-hop circles—is being able to create a standout groove. In rock and pop styles, the groove may come from just the drum and percussion parts, but for some styles, the beats, loops, and sam-ples make up the majority of the track. This means that, as a producer, you'll need to have your workshop together, and be able to put together tracks quickly and easily.

With programs like Reason, Logic, and GarageBand (and their libraries of Apple Loops), Cubase, Pro Tools, Ableton Live, Fruity Loops, and BFD, the sounds and se-quencers for putting your track together are all there; you just need to make them work for you. The problem with PC/Mac-based sequencers is that you'll go insane "mousing" around all over the place to get a rhythm track together. This is why most engineering producers work (at least partially) on a few important pieces of hardware as well:

■ *Keyboard Controller.* The bare minimum to accompany any sequencing program is a MIDI controller. If you're not a keyboard player, just get yourself a little two-octave controller; it's still useful for programming drum patterns and controlling parameters in the sequencer. For less than $100, you can get an M-Audio Oxygen, Korg Nano Key, or something similar. Even if you're working on guitar-driven rock, you'll find a controller irreplaceable for creating arrangements when demoing or producing. Those of you who also play keyboard instruments may want to invest a bit more in a weighted, after-touch-sensitive controller with plenty of controller knobs and faders for easy control of synth and effects parameters.

■ *Pattern-Based Sequencer.* Also called a step sequencer, these units display rhythm patterns in a very intuitive way, making composition and editing simple. Figure 9.1 shows Reason's ReDrum as an example. After choosing a set of sounds for kick, snare, hi-hat, and so on, each drum's pattern is then shown on a series (usually 16) of backlit pads. If each key represents a 16th note, then one complete bar is shown on all the pads. Drum patterns are easily created, copied, pasted, and modified, allowing you to create a whole rhythm track arrangement in no time.

Figure 9.1 Reason's ReDrum is a software version of a step-based sequencer. Although it's easy enough to click on the switches and pads, having a hardware controller makes programming drums even more convenient.

From within your software (or hardware) sampler, you can get your dirty little hands on an endless variety of drum samples, from classic drum machine samples (think Roland 808), to immaculately recorded drum kits, funk-filthy sounds, and modern layered synth sounds. Most modern producers layer kick and snare sounds, and this is also easily done with a step sequencer. You just copy the pattern, assign it to a new sound channel, and flip through your library until you hear what you want. Reason allows you to let the pattern run and listen as you flip through your sounds—a big time saver.

■ *Hardware Sampler/Sequencer.* The classic is the Roger Linn–designed Akai MPC60. It's probably the most well-known and loved bits of production gear ever manufactured. This unit (or perhaps a later model) is a fixture in the project studio of just about any noteworthy producer. The ability to quickly load sounds and cut samples, combined with trigger pads for programming patterns, makes the MPC60 one of the best ways available to create a basic rhythm track.

Many years ago, while setting up producer Danny Kortchmar's project studio, I popped a disk into the MPC60 to test the output, and was hit with the unmistakable rhythm track to Don Henley's "Boys of Summer." I was surprised to hear this coming out of what I, at the time, thought was really just a toy for making demos. It turns out that Kortchmar, among many, many others, uses the MPC60 to create the tracks that wind up on these hit records. Just owning a great sampler/sequencer won't guarantee you success—you need to bring some great ideas into play, and this comes from lots of listening, imitating, and experimenting—but the right tool for the job certainly helps.

When working on a track, it's also great to have some hardware that anyone can just hit and get a sound out of it. Unlike a computer, with its inherent mouse-dependency and endless menus, a unit like the MPC60 lets everyone get their hands into the process of programming a beat. One thing that I can't stand is having to click about on a computer while everyone waits to hear something. With hardware interfaces, both producer and musicians can get their grubby hands on the music and punch out ideas quickly.

Production and Generative Music

Considering the unprecedented volume of music being created today by the multitudes with GarageBand, Reason, and the rest, the next logical step is to involve the computer itself in the creation, variation, and inspiration of new music.

People have a tendency toward organization and pattern-based thinking, which is necessary for us to be able to make use of the flood of sensory input the world provides. People tend to find order in chaos, and inspiration in randomness and disorganization. Granted, sometimes we create our own chaos, although those who swear by their disorganization are likely to really have a method to their madness, which is just a variation of organization.

It's difficult for people to be random in thought or actions; it can even be difficult to let go of certain patterns of behavior. Just think of all the exercises, methods, drugs, programs, groups, and therapies that people undertake to try to break habits! You may be familiar with free improvisation, stream of consciousness, meditation, or psychedelic substances; things humans experiment with in order to try to "free the mind" and gain an immediacy of experience.

Computers, on the other hand, are pretty good at being random. In many cases, it takes a lot more work to create a reliable way to organize input on a computer than it does to create an algorithm that causes a computer to generate randomized output. (Okay, mathematicians, I am generalizing a bit, so for you I'll say "apparent randomness.")

Generative music programs allow us to delegate creative processes to a computer. Although you might normally rely on your computer simply for cold number-crunching, why not let your computer run with your ideas? You can then sit back and enjoy the variations, or even take on the organizational task of sifting through what you hear for inspired moments.

Generative music programs are *not* just generators of random bleeps and tones. Keys, scales, rhythms, and harmonies can be defined as limits, and complex rules created to keep things from sounding random or atonal. Ultimately, you set the parameters, but then the computer begins running with the ideas.

Here is a quote from Brian Eno on the subject:

> *Until 100 years ago, every musical event was unique: music was ephemeral and unrepeatable and even classical scoring couldn't guarantee precise duplication. Then came the gramophone record, which captured particular performances and made it possible to hear them identically over and over again.*
>
> *But now there are three alternatives: live music, recorded music, and generative music. Generative music enjoys some of the benefits of both its ancestors. Like live music it is always different. Like recorded music it is free of time-and-place limitations—you can hear it when and where you want. I really think it is possible that our grandchildren will look at us in wonder and say: 'you mean you used to listen to exactly the same thing over and over again?'*
>
> —Brian Eno, 1996

Artist/Producer Eno has made use of generative music programs both for composition (of what some consider to be esoteric music) as well as to seed the creative process for his productions of popular artists (Bowie, U2, and James). Generative music adds an element of unpredictability and surprise, like an additional composer and performer. It is not fixed or repeating, but endlessly varying, based on your initial input. I highly recommend that you investigate the possibilities—it may not be your grandchildren who wonder why recorded music was listened to the same every time.

At the time of writing, the first generative music programs have begun appearing as iPhone applications, which will expose multitudes of users to generative music programs. It's just a matter of time (weeks, I'll bet) before one of these programs make their way into music that sweeps across the web on YouTube or some social network, opening the doors for generative music on a new level.

If this interests you, I recommend trying the program Noatikl from http://www. intermorphic.com/, shown in Figure 9.2. You can use it for free for 30 days, and it's otherwise very affordable. Brilliantly, Pete Cole and Tim Cole (Intermorphic's wizards behind the curtain) have programmed plug-in versions of Noatikl for Logic, Sonar, Cakewalk, and GarageBand. You can also use the stand-alone Noatikl program to communicate via MIDI with any program (for you Cubase and Pro Tools people).

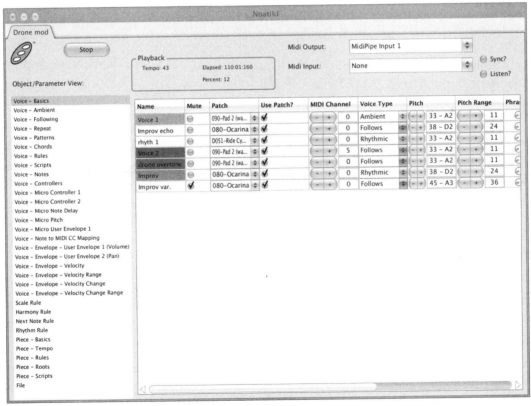

Figure 9.2   A window in the Noatikl v1.5 interface. Each MIDI voice is adjustable by numerous parameters that determine its behavior. Users determine how freely or strictly Noatikl may improvise, voice by voice.

This means that the elements of "controlled randomness" that generative music creates can be easily integrated into any music project. You can use the MIDI signal to play synths, vary fills in MIDI drum and percussion parts, develop melodic patterns, trigger audio loops and samples, even control mix parameters like reverb depth, panning, filter frequencies—*anything*. You can invite a fresh wind into your carefully constructed castle of audio. Sounds like a concept—"controlled randomness" rings of "perfect imperfection."

Noatikl works by setting up individual "voices," which are then assigned rules as to how they play. Voices can be set as rhythmic, ambient, following another voice, playing fixed patterns, repeating (with variation), and even "listening"—waiting for MIDI input to react to.

Users then determine scales, keys, harmonies, how probable it is that certain notes may be selected, how often and in what note values the voice will play (or if the notes should even fall into measure boundaries), and a spectrum of other aspects of performance. By using the "listening" setting, HAL—oops, I mean Noatikl—can be asked to improvise along with you, adjust MIDI parameters while you play, and harmonize to your performance.

Although it appears overwhelming at first, you can gain a working understanding quickly by following the step-by-step tutorial for creating a first session. At that point, natural curiosity tends to take over, and by adjusting settings, you learn very quickly how Noatikl reacts to settings.

At first glance, it may seem that Noatikl would be most suited to soundtracks or ambient music. However, there are some very inspiring drum templates available. Generative music could bring subtle variations to the typical programmed patterns and loops DAW engineers tend to work with. I can't imagine a better fix for bored, tired ears after hearing a song 487 times during tracking, overdubbing, and mixing. Naturally, you can record the MIDI of every take Noatikl performs in your DAW (in case you hear that one magic take); MIDI takes up very little processing power and hard drive space.

Ableton's Live program is another option for adding generative music-style unpredictability into your music. Although not as in-depth in terms of the "generative" options as Noatikl, Live may be more useful to you if you're looking to vary certain elements in an arrangement.

By creating rules and conditions for related audio loops, MIDI notes, and more, you can instruct Live to play back these elements based on chance (percentage set by user), rules (if this, then this), and the conditions that follow.

Imagine that you have a basic drum loop playing throughout a song. It is likely that you want to have some sort of drum fill occur, say, every eight bars. By manipulating these rules and conditions, you can have Live select from a range of fills, or perhaps play no fill at all, and just drop the loop out for a bar. Combine this with a variety of percussion parts, varying high-hat patterns, and accent sounds, and you can create a rhythm track that provides you with more than a repetitive, programmed pattern.

This makes an excellent tool to which you can practice an instrument or work on writing a song; providing you with more variation than a set CD practice track or an inflexible basic track to which you're trying to compose lyrics. For those of you who are more adventurous, Live can bring an element of interest and surprise into live performances to backing tracks, whether the whole song or just select sections for improvisation. Having an element of unpredictability to react to feels much more like having real musicians there with you—varying the percussion fills, forgetting chord changes mid-song, and dropping drumsticks in a drunken stupor. *Viva la Tecnología!*

Final Thoughts

Overall, this chapter on production probably sounded somewhat philosophical. That's kind of the point. Production is not any one particular method or technology discussed in this book, but the interaction of any and all of them with the human side of the story. Just as music can't be boiled down to "just a beat with some chords and a melody on top," production is not just the sum of technique, tracking, and mixing, but a process that involves all these steps in a complex way.

10 Building Your Arrangement Chops

Sometimes producers also take part in arranging the songs. If you are self-producing your tracks, this is naturally the case, but in some cases the songwriter may be open to arrangement suggestions. Arrangement is not an exact science; there are no set arrangement rules. You will need to use experimentation and experience to determine what works best for every song.

I suggest that anyone interested in pop arrangements read the excellent analyses of The Beatles recordings by Alan W. Pollack. They can currently be read online at http://www .icce.rug.nl/~soundscapes/DATABASES/AWP/awp-notes_on.shtml.

If that link is out of date, just search for "Beatles Alan Pollack analysis," and you will find them. If he decides to take them offline and publish them in printed form, just trust me and buy the book—it is a true goldmine of arrangement ideas.

Even if you aren't a Beatles fan, the collaborative genius of Lennon, McCartney, Harrison, Starr, producer George Martin, and engineers such as Geoff Emerick brought about undoubtedly the best arranged pop/rock songs of all time. They have influenced everything you hear today. The long arrangements of progressive and metal bands can even be seen as a counter-movement, which is *still* a type of influence!

In fact, if you dislike The Beatles, then you should *know your enemy*. As a good producer in any style, ignorance is not bliss. If you want to avoid a Beatle-esque, pop sound, you should eschew even vocal doubling, consciously avoiding that particular vocal sound. A true punk band, for instance, consciously avoids "slick" arrangements and studio tricks. Keep in mind, however, that modern high-profile "heavy" productions (Slipknot, System of a Down, Metallica, and Children of Bodum, as examples) make use of many of the standard tricks, just fit into their style; doubling, harmonies, keyboard pads—they're in there! The same goes for hip-hop and electronic styles. Vocal doubling, harmonies, background parts, song section arrangement, copping old blues riffs; these elements are all part of the music tradition. Be conscious of them, and question what you use as your "standard" production moves.

The way to begin learning how to arrange music is by listening intently to particular parts of the music. This takes some practice and lots of concentration. Most listeners don't really hear the parts of a song arrangement as separate parts working together; they just hear "music." You're undoubtedly aware of the separate parts, but try this— ask a few different people to listen to a song with you (at different times, so that they don't say, "I hear what you mean" to go along), and ask them to listen to just the snare drum, bass guitar, piano part, or the like. Many people hear the vocal as separate, but the music as "that other stuff." You may even have to explain that the "drums" are different pieces, and find it difficult to point out one piece while listening! Consciously listening to music is something that you should practice, and it makes listening more enjoyable. Just because you can hear the separate parts doesn't mean that you can't let the whole impression wash over you. With a little practice, you'll be able to hear a song in many different ways, and get a lot more out of it.

So be sure you can pick out a string line, vocal harmony, or percussion part by ear. Listen carefully to your favorite artists, and consider where these types of parts are used, and where they aren't. Figure out how you could use a vocal harmony, for example, in a song of yours that's in a different key and tempo. Perhaps that part is a sustained single note a fourth below the lead vocal that lends the chorus a hypnotic quality that you dig. If you know the concept, you can translate it to another situation. Then it's up to you to try it out and see if you like the results.

Food for Thought: Song Parts/Sections   It is important to understand the different parts of a song, and develop a lingo with your partners in musical crime. It will help you all easily refer to the various parts of the song. Whether writing, rehearsing, recording, or mixing; this is good practice. Here's a list of terms:

Intro: The beginning. Duh. This may be as simple as a two-chord change that sets the key, or it may draw on material from another part of the song. Some ideas for introductions are found later in this chapter.

Verse: The basic section of any song, wherein the lyrics tell the story. In instrumental music, this would be the basic melodic theme. In folk or traditional music, the song may consist only of multiple verses with a similar ending. More common in pop/rock is a verse section that is followed by a chorus/refrain section. Common arrangement shorthand is to call the verse the "A section."

Chorus: The "B section" of a song, usually presenting a memorable lyric and melodic hook. Lyrically, the chorus provides a central idea that relates to the story told in the verses. Clever lyric writers can make the same chorus take on a different meaning as the verse's story progresses. Check out Sting's "I Hung My Head," which lyricist legend Johnny Cash also covered. The song's protagonist first hangs his head in shock, then in remorse, and in the end is hung with a rope.

A brief note about lyrics: Needless to say, Johnny Cash's lyrics make for a great study in songwriting and structure. Other great phrase-turners you should be familiar with include Bob Dylan, Leiber and Stoller (many hits of the 50s), Holland Dozier and Holland (Motown), Tom Waits, Simon and Garfunkel, Chuck Berry, Lennon and McCartney, Jagger and Richards, Buddy Holly, James Brown, Run DMC, Ice Cube, Stevie Wonder, Prince, Elton John, Neil Young, and Lou Reed. This is by no means a complete list, but one thing's for sure; every one of these writers has had a hit song, and it's the lyrics that do it every time. This reinforces the concept of "song first" in the list of priorities.

Pre-chorus: Sometimes putting the verse right up against the chorus just doesn't work. The listener might need a moment to consider the lyrics, or the music might need a dynamic or harmonic transition. In this case, a short section can be added to lead from the A section into the B section. Sometimes this section doesn't lead to the chorus the first time around, but goes back to a verse. Later in the song, you can leave out the pre-chorus, again creating a surprise for the listeners.

Bridge: This is the "C section" of a song, and normally uses a new melody and chord progression. It provides variety, a release, and/or a solo section. A change of key is also typical, and writing an effective bridge may require a bit of music theory knowledge, although the way it sounds is always most important! Sometimes the bridge is placed in the arrangement instead of a chorus after a verse. This, again, provides a surprise for the listeners, and can be a great way to revive interest in a song if the number of verses makes the song particularly long. If a bridge has lyrics, they're usually different than the verse and chorus, providing a different angle on the story.

Tag/Coda/Turnaround: A few ways of describing a similar part, depending on where it shows up in the song. This is the flip side of an intro; a short section that usually follows the chorus, bringing the song back around so the verse can start again. It's not uncommon to use the tag as an intro; U2's "Vertigo" is a perfect example—that fuzzed-out guitar part that starts the song comes back after the chorus, creating symmetry in the song before the next verse starts. The term "turnaround" refers to a chord progression that leads back to the opening chord of the verse. Progressions like "IV, V, I,"; "V, IV, iv, I,"; and "ii, V, I" are typical turnarounds. Knowing what is typical lets the clever producer manipulate the listener's expectations.

Outro: The ending. Repeating the chorus an extra time, and ending on the tonic chord is a typical move, as is the old "repeat and fade" idea. Fading on repeats of the chorus can be made more interesting by introducing a new countermelody. In the style of a turnaround, using a small part of a chord progression from another part of the song will sound familiar while still providing a new part to end on. It's less common to hear entirely new material as an outro, although in progressive/experimental styles of music it is often done. Keep in mind that a concise ending to a song adds to the "play it again!" factor. Unless you have an especially great melody that bears repeating, avoid the long, repeating, fade-out.

Solo: This is where the instrument players get a chance to stretch out and work with the melody. In folk, country, and sometimes even in pop music, the solo melody is often similar to the vocal melody, with just a few embellishments. The solo section may use the verse progression, the chorus, and verse and chorus, or it may be an entirely new section with a new chord progression and even in a new key. Sometimes a flashy solo works, other times it is too much for the listeners. There are plenty of songs in which a complex solo fits well to the pacing of the piece, keeping the listeners interested. Try to imagine "Comfortably Numb," "Sultans of Swing," or Hendrix's version of "All Along The Watchtower" without the solo! However, when in doubt, simple is often best. If you're not familiar with Neil Young's solo in "Cinnamon Girl," have a listen. Consider this concept for your music. Pure attitude can be so much more powerful than the notes themselves.

Break: Almost like a "subtractive solo," a separate section can be created by dropping the music arrangement down to fewer elements, often just the percussion. The "drum break" is tried and true, but may be cliché. You need to make the call. Consider a different type of break! You may have a supporting part, such as the strings, that's mixed rather low although it sounds great. Try dropping everything else and turning that part up. The end of The Verve's "Lucky Man" does just this. "E" of Eels is also fond of the unusual break. Give this a try—it's so easy to do with DAW mix automation; just use the Solo switch, and have a listen.

Vamp: A basic chord progression played repeatedly by the rhythm section while they wait for a cue from the lead player/vocalist. This can show up at the beginning or end of a song, or may lay the foundation for a solo section. If a vamp section is noted on an arrangement sheet, the players will be paying attention, expecting a cue to be given to move into the following section.

Understanding Song Structure

Most songs consist of separate sections, even if the differences are subtle. A pop song can have a glaringly obvious section arrangement such as:

Intro > Verse 1 > V2 > Chorus > V3 > Chorus > Bridge > Chorus > Outro

Whereas a minimalist electronic music piece (Kraftwerk or Paul Kalkbrenner) may have a structure more like a classical music "theme and variations" piece. The similarities can be striking; there are often *ostinato* (repeating) basslines, a basic melody that repeats often and over changing accompaniment, and changes to the theme during the piece, eventually returning to the original melody form.

I certainly caught up on a lot of sleep in the required classical music courses at college (Sorry, Ralph and Al!), but in retrospect, an understanding of the "Sonata Allegro" form gave me a lot of insight into how music can be effectively arranged.

The Sonata Allegro form (also just called Sonata) is the standard form of classical symphonies, and is based on developing a theme. Pop, rock, hip-hop, whatever-you-got;

there's *always* a hook of some sort—that's your theme. There are many ways of doing this, but let's consider the basic Sonata structure:

Introduction > Exposition > Development > Recapitulation > Coda

- *Introduction:* No problem—that's an obvious one. Just about every song has one. Even if it's just a drum fill, something introduces the song. Typical in the Sonata form, as in pop music, is the use of melodic ideas from elsewhere in the piece, and possibly a slower tempo or even free time. Think of The Beatles' "Taxman;" a collage of sound wherein a voice even counts off the song in a different tempo.

- *Exposition:* Here, the main melodies (themes) of the piece are presented and usually repeated to establish them for the listeners. To make a pop music comparison, the exposition could be likened to the verse, although because the classical form is generally much longer, this section will probably sound complete in itself. To the casual listener of classical music, the exposition is the part you identify as the piece, such as Mozart's "Eine Kleine Nachtmusik."

- *Development:* To continue the pop comparison, this is similar to the bridge or solo. This is the part of the piece in which the basic melody is changed, twisted, turned, developed, and ultimately set up for the return to the original version. My former professors would kill me for simplifying like this, but they can't fail me anymore, so there.

- *Recapitulation:* This makes sense—"recap"—the return to the original theme/ melody. Although the way the repeat is presented varies from piece to piece, it's important to reinforce the theme/melody with the listeners. If you listen carefully, almost all music does this, even if the melody is dressed up differently when it returns. Those of you looking to break rules should consider this! If you don't at least hint at a melody from earlier in the song, the listeners could feel that the song just rambles along and ends somewhere. Not that this can't work—you just might need a different strategy, such as in a through-composed song like "Stairway to Heaven."

- *Coda:* Put simply, this is the end of the piece, the outro, the final cadence or "turn-around," in pop music terms. I'd dare say that classical music endings tend to be drawn out, but it's all a matter of perspective. A recap of important melodies along with a satisfying cadence fits a longer composition better than it does a short pop song. Everything comes to an end, if you'll pardon the philosophical note, but in pop music you may find a double chorus, a repeat of the turnaround (especially in blues), an outro vamp, a breakdown, or even a chaotic "falling apart" effect, as in some Beck songs. The old "fade-out on the chorus" is so overdone by now, that writers, producers, and arrangers are always looking for something different. Do your listeners a favor, and come up with a cool idea for the ending, or just keep it simple!

All in all, the Sonata form is very much like the classic *AABA form,* an arrangement that shows up often in pop, folk, soul, funk, and jazz styles. Typically, George Gershwin songs are named as examples, but younger readers may be more familiar with George Harrison's "Something," or Bill Withers' "Ain't No Sunshine (When She's Gone)."

"Ain't No Sunshine (When She's Gone)" is a perfect example of the AABA form. The verses (A sections) are different lyrically, but all end with the hook line, " . . . any time she goes away." The B part is the repeats of "I know, I know . . . ," which Withers originally intended as a placeholder over which he would later compose a "real" B section. Well, it turns out that the "I knows" stuck, and the song is a perfect example of an AABA form with an outro repeating "Anytime she goes away," which drives that hook line home. A simple, effective song.

Taking the AABA form a step further, and returning to the B section again makes the form AABABA. The Beatles' "A Hard Day's Night" is a perfect example of this form. As an intro, Harrison compiled an extended chord out of notes from all the chords in the A section (major chords G, C, F, and D), creating a G13add4 chord (G, C, F, D, E). The 12-string guitar and overdriven delay effect create an unmistakable, if very short, intro. This returns as the outro, although performed differently.

More common in folk songs, the simple *AAA form* is worth a mention. It's just a series of verses with different lyrics but the same melody. The AABA form is just an extension of this. Sometimes a bridge is unnecessary, and only interrupts the story of the song. It's always worth considering this simple form. Bob Dylan's "Blowin' In The Wind" is a classic example of this form; an ongoing story that needn't be interrupted by a change of chords and melody. This is as simple as it gets!

The *verse/chorus form* is probably the most common song form you'll hear. Just turn on the radio, and chances are that'll be the structure being used in the latest hit. With added bridges, pre-choruses, tags, solos, and breakdowns, this form is often expanded, becoming more complex, but is still based on the idea of a separate verse and chorus. Key to this form is that the chorus delivers. The verse is the story, and the chorus a high point; the hook. In the AABA form, the verse has a hook line that you can latch on to, but it isn't as extreme and separated as in the verse/chorus structure. Also, when adding a bridge to the verse/chorus form, it really becomes clear that the verse and chorus are separate entities; the bridge often leads straight back into a chorus, and you realize that they are clearly separate sections.

Now that you have an overview of some of the basic song forms, you can better communicate during the songwriting/rehearsing/recording process. "We need to redo the vocal harmony in the second 'A' section," or, "something is missing in the bridge; can we try adding a second guitar part?" lets everyone know just what you're referring to.

When working with a session musician, it is also good practice to refer to sections of the song in these terms, as well as to provide a chart of the chords with the sections clearly marked as A, A, B, A or Intro, V1, Chorus, V2, Chorus, Bridge, Chorus, Outro. Any player with a bit of session experience will expect this, and it will save you time, money, and communication frustration!

Thoughts on song structure:

- *Remember that the song dictates the form, not the other way around.* It's rare to start with a form and fill in the blanks. Generally, the musical ideas start flowing, and then you look at it and say, "Aha! That's a verse/chorus structure, with a pre-chorus and a bridge." However, if it feels like the song is missing something, considering some of the typical forms can help you come up with ideas as to what can be added. Most songs start out as a simple melody and chord change, or sometimes two pieces of music that fit together in melody and lyric, tempo, and feel.

  If the lyric is strong enough, leaving these parts as they are, and using a simple AAA form may be enough. Create a simple intro, perhaps just the last four bars of the verse played instrumentally, and you're good to go.

  But just because a song is simple doesn't mean that it's perfect. On the flip side, a lyric that isn't strong enough to work in an AAA form doesn't mean that the song is a loss.

- *If a song is not working in its original form, try adapting it to another form.* An AABA song could have the B cut out, and become AAA. Maybe the B section was unnecessary, and this will be perfect. It could also be that verse/chorus is too "over the top" for a simple song, and the chorus should be simplified. Maybe shortening the chorus and joining it to the verse to create a simpler A section, and then adding a B section to make an AABA form will do it. A friend of mine did this for a song of mine once, removing a "top of the mountain" chorus idea that in retrospect was pretty lame. The song became a simple AABA, now that I think about it, and works well now.

- *By adding a B section or a bridge, the impression of a song can change.* The new chord progression and melody of a B section/bridge can refresh the ears, giving the following verse more impact. New lyrics in a B section/bridge can provide a new perspective on the verse lyrics, perhaps adding a twist to the story that improves the song. The right intro can put a stamp of originality on the song, grabbing listeners' attention immediately. Adding a tag after the chorus can create "room to breathe" in the song, helping the lyrics come across clearly.

On the other hand, a song weighed down by too many turnarounds, pre-choruses, solos, and other trickery will quickly lose focus and sound overproduced. You need to find the balance.

■ *These forms are not hard and fast rules!* MGMT's "Time To Pretend" uses a verse/ pre-chorus/chorus form, repeated twice—ABC ABC. By dropping out the drums in the first half of the second verse section, the song gains a dynamic that the heavily compressed track would otherwise lack. The chorus section melody is used as the intro, making it show up at the beginning, middle, and end of the song. Everyone who hears the song once will recognize that melody, and know what's coming when they hear it again.

As mentioned earlier, Led Zeppelin's "Stairway to Heaven" is unusual in that it continually adds new sections. It's sort of leap-frogging, in the manner of ABA BCB CDC. Although "Stairway" is more complex than that, the idea of progressively adding new sections without returning to the original section at the end of the song is certainly possible. A good basic concept for this form, "one step forward, one step back, two steps forward, one step back" (as ABA BCB CDC, and so on), maintains a forward drive while keeping things familiar.

Considering Song Sections Up Close

When arranging the song structure, you have a unique power over the impression of the song, in terms of how the whole story unfolds. Moving in closer, each section is composed of numerous instrumental and vocal parts that define it.

It is rare that any given section (verse, chorus, tag, and so on) repeats exactly when it returns in the song. Even in songs where the second verse is a lyrical repeat of the first (and this does happen), there will be some defining elements to set the repeat apart. The singer may alter the melody, the drummer may *tacet* (stop playing for a period), strings may enter, percussion parts may change; there are literally hundreds of possibilities. These parts and possibilities are part of the arranger/producer's bag of tricks.

The following sections contain some arrangement ideas for each part of a song. If in doubt, always focus on and serve the lyric, occasionally leaving the listeners time to consider what they've heard. Sometimes this means just hanging on one chord for an extra couple of bars.

Introduction

Spend some time listening to what other artists use as song introductions. They can be quick and in-your-face or compositions in themselves. In any case, a good intro is un-ique and recognizable. There's no formula for a good intro, as that would defeat the purpose of unique, but there are a few ideas that you can use to get your creative juices flowing:

- *In rock and pop songs, the intro is often an instrumental version of the verse or chorus, shortened and with one instrument playing a memorable hook part.* As noted, MGMT's "Time To Pretend" is a perfect example. So, for a hook intro, try using a bit from the chorus. If you don't have an instrumental melody/counter-melody for the chorus, create one. Arpeggiate the chords of the chorus progression, compose a simple synth, string, or piano countermelody, or just start with the chord that resolves to the verse, and then groove on the verse for a bit before the vocal enters.

- *Another great way to borrow material from your own song is to sample a section from a verse/chorus/bridge and process it to alter the sound.* This creates a more oblique reference to a later part of the song. Try running the sampled part through a reverb and using only the reverberant signal. Filtering and compression can create a classic "coming through the radio" sound, which may have been used often, but still makes a great addition to a sound collage. Also consider playing the same part on a different instrument; adapt a piano part for guitar, whistle a melody, play the hook line on bass guitar, imitate the drum break on muted acoustic guitar—get creative and switch things about!

- *Sometimes a part unrelated to the verse or chorus shows up as an intro.* Often, this will have something to do with the bridge, but may be completely unique within the song. John Frusciante (guitarist of The Red Hot Chili Peppers) is a master at creating intros. "Under the Bridge" is a great example, and it's not even in the same key as the song! This can be a great place to use a song fragment that never came to fruition. Most writers have some kind of little ditty that they never manage to turn into a whole song—record a bit of that and try putting it in as an intro.

- *The guitar intro is still the standard for rock tunes, and a great intro can make a song legendary.* Think of all the great players who've based careers on smokin' guitar intros—Jimi Hendrix, George Harrison, Keith Richards, Brian May, Jimmy Page, Angus Young, Mark Knopfler, Eddie Van Halen, Alex Lifeson, Slash, Jack White, the list goes on. It's amazing how versatile the guitar is, and how all these guitarists manage to be identifiable within just a few notes. If you dare step up and create a guitar intro, do it confidently. Whatever tone you choose, push the limits. If you want smooth sounding, make it pure butter. If you want "off-the-cuff," record it just once and don't polish the rough edges. If you want aggressive, make it howl and bleed.

- *I admit that, as a guitarist, I am partial to guitar intros. However, every instrument can step up for a virtuosic flourish, a tasteful cycle of chords, or a butt-kickin' groove to get a song going.* Borrow from your own influences, combine, meld, and interpret that which you dig. Consider piano greats Jerry Lee Lewis ("Great Balls of Fire"), Nicky Hopkins (Rolling Stones "Loving Cup"), Tom Waits' simple folk/traditional intros, or the borderline-insane playing of Mike Garson (David Bowie). Drummers can show off a slick fill, or consider grooving song starts along the lines of "When the Levee Breaks," "Walk this Way," "Paint It Black" (after the sitar line), or even the painfully simple but effective "Billie Jean" intro. And let's not forget the strings that start off The Verve's "Bitter Sweet Symphony;" a great part, but it crossed the line from borrowing into plagiarism, so beware!

Verse

Because the verse is usually one of the initial parts composed for a song, it's often clear which parts are needed to support the vocal. Most often, the verse is the section with the fewest additional instrumental parts. In many cases just the vocal and an accompaniment on a chordal instrument (piano, guitar, accordion, and so on) will do. If you've got drums and bass in there as well, make sure your groove is really on the money—if that's not happening, nothing else will work.

The verse is the *story* of the song, so give right of way to the lyric, especially in the first verse. If you have some great backing part ideas, save them for later verses. As listeners become familiar with the groove and the melody, those basic parts begin to fade into the background. This is the time to introduce more action into the verses, and this also builds the dynamic of the song. The opposite is also true; removing parts that have become "familiar" over the course of the song grabs attention.

Some specific ideas to experiment with:

■ *The timbre of the vocal is presented in the first verse.* This means that your mic choice and how you choose to process the vocal makes a big impression. Consider the different ways a vocal can sound: smooth or brash, reverberant or dry, filtered or full-range, single- or double-tracked, small or large, lazy or energetic, serious or sarcastic. Play with these contrasts and consider where they may change within the song. Switching from one to another can be incredibly effective. A great example is Beck's "Information;" the verse vocal is double-tracked and rather lazy sounding. When the chorus enters, the vocal is double-tracked, and the performance energetic; a brilliant arrangement.

■ *Usually the writer's main instrument winds up as the main supporting instrument in the verses.* If the magic is not happening somehow, consider a switch. Swap acoustic piano sounds for synths or guitars. Give the strummy-dummy acoustic guitars a break and try simple block chords. Do away with syrupy synth patches, and try stacking up real vocals—it's easy to do—just track one note of the chord changes at a time, track for track. Consider humming, "Oohs," "Ahhs," or just the beginning of words from the lead vocal. If the lyric was, "How could you dream of meeting me," you could vocalize on "haah-," "dreeah-," "meee-," for example.

- *If things sound cluttered in the verse (too much competition for the vocal), even though you haven't layered in a lot of parts, consider the rhythm of the parts.* Where a part enters can make all the difference in the world. Bass and drums are entering on beat one, so try having keys, guitar, background vocals, and so on come in on the "and" of one (the second eighth note of the bar), or even later. If everyone is coming in on the downbeat, there could be a traffic jam. I've had great luck with background vocals, for instance, by just sliding the whole track back by an eighth note in the DAW edit window!

- *You may want to keep the dynamic up after the first chorus, keep the groove going, especially if the song should be danceable.* A great example is Peter Gabriel's "Sledgehammer." The intro lets us know what's in the bag of tricks, but the first verse is broken down to the basics. Verse two keeps rolling along after the chorus with horns and vocal harmonies, and the verse after the bridge adds background vocals echoing the lyrics. Anyone interested in great arrangements that can be pulled off live should spend some time listening to Gabriel's "Secret World Live" album.

- *Don't forget percussion!* Shakers, tambourine, triangle, finger snaps, maracas, and other small percussion instruments and loops give a track momentum. They "blend away" into the listener's ear very quickly, becoming part of the collective background music. By bringing these parts in and out of the mix, the song's dynamic can be improved. Let's say that the song begins with drums, bass, and a percussive loop. Piano chords and the vocal enter together in the first verse. At the beginning of the second verse, dropping out the percussive loop brings more attention to the vocal. The loop, which we're used to hearing anyway, could resume a couple bars later, giving the song a "push" without being distracting. Add a tambourine or shaker in the chorus, and the dynamic increases again. Check on yourself once in a while, though—mute the parts you've added, and make sure it wasn't actually better when simpler.

Pre-Chorus

If your song has a pre-chorus, you'll want this section to stand out from the verse; it should be clear that the song is headed somewhere else. Often, simple touches help build a pre-chorus, it's important not to get too big yet, as that could take away from the impact of the chorus. A few simple ideas:

- *Start the vocal doubling here, or otherwise change the vocal tone.*

■ *Simple accents on the chord changes make the section stand out.* Pitched percussion, an additional instrument playing the chords, or a new part doubling another instrument are tried-and-true pre-chorus ideas.

■ *Subtle synth pads/noise/ambience give a sense of heading somewhere.*

■ *The choice of chord inversions affects the restlessness of the music.* Root position chords are most stable, whereas inverted chords (a note other than the root in the bass position) create tension. By using inversions, it's easy to make the pre-chorus more tense, even if you haven't changed the chord progression radically from the verse. Try having the bass player play thirds or fifths in the bass, or create an ascending or descending bassline, even if some of the notes sound very tense. Our ears accept harmonic tension very well if it is part of a clear motion.

■ *The drum part is especially important in the pre-chorus.* Let's say that the verse is played on the hi-hat, and the chorus on the ride. As a way of bridging the gap, the drummer should be opening up on the hi-hat, adding tom fills or cymbal accents, or at least altering the kick/snare pattern that was the verse. If you're programming drum parts, keep these tips in mind; don't just roll through the pre-chorus with the same part as the verse.

Chorus

Here's the payoff of the song, the part everyone should be singing in the shower and cranking up when it comes on the radio. There's always some part that makes the chorus stand out, even in sparse arrangements.

The obvious candidates are vocal harmonies, background vocals, strings and synth pads, distorted guitar chords, ride and crash cymbals, horn section stabs, and all manner of tambourines, congas, triangles, and other percussion accents.

Sometimes the usual suspects are just what you need. On the other hand, doing the expected can really take the steam out of the engine if it makes the song sound clichéd. What can a budding arranger do to avoid floundering in a sea of sameness? Once you know the typical moves, play around with expectations; mix the usual with the unusual. Consider these points:

■ *When working on vocal harmonies, try something other than the typical thirds harmony that has become the default.* Try adding a voice singing on just one note through the whole chorus. There are even plug-ins that can force this. This will create unusual harmonies against the lead vocal. If you do the usual harmonies, try manipulating them with effects like rotary speakers, odd reverb, amp simulators, and pitched delays.

- *If you want to beef up the sound of the chorus without adding a big wall of fuzzy guitars, synths, and background vocals, try having the bassist overdub a fuzzed-out duplicate of the bass part.* Add one extra guitar track, doubling the rhythm part, panned hard to one side and filtered. Without sounding overproduced, the track will suddenly take on weight and impact.

- *Manipulate the drum sound in the chorus.* Layer the snare with samples, treat with effects, and experiment with panning to create a wide sound. Automate the room mics (or add a good room reverb) to add more depth and size to the kit. Try adding an exciter effect to an aux send and pushing that up as another way to make the sound jump out.

- *Acoustic guitars are great for building up a chorus, their percussive element adds as much as the chords do.* After tracking the basic chords in open position, slap a capo on the neck, and playing the chorus progression up higher on the neck. The combination of chord inversions and layered strings sounds like something between a standard guitar and a 12-string. Another method is to set up a Nashville-tuned guitar. By replacing the low strings on a standard six-string guitar with thinner strings, and tuning them one octave higher, you get the higher strings of a 12-string. Now double the standard six-string part with this guitar, and you're in jangling, shimmering guitar heaven.

- *Add odd noises and samples.* Anything that repeats a few times eventually becomes an accepted part of the sound collage. When you're using the chorus without vocal as an intro—a common and effective method—these sizzles, blurps, wah-wahs, and bleeps really stand out. When the chorus comes in, they sound familiar, and tend to make sense, no matter how far out they are. Try it, you may be surprised what you can get away with!

Bridge

A true bridge will introduce a new melody and chord progression, already distinguishing the section from the rest of the song. It's tempting to experiment with new sounds and parts in the bridge; just be sure they serve the song. If you are working in a style that should be easily performed live (folk, country, and blues), be sure to stick to the instruments already in the arrangement! It could be difficult to come up with steel drums and harpsichord on the spot at a barn-raising festival.

If you're working in electronic styles (hip-hop, house, trance, techno, and so on) then let 'er rip! The bridge/solo instrumental section is the place to pull out all the stops. Try to maintain something from the body of the song as a thread through the section (like a bass motif, drum pattern, lyrical connection, loop/sample, and so on) to keep the bridge from sounding entirely disconnected. A clever arranger can make a radical change at the start of the bridge, working their way back into the chorus by the end of the bridge. In this case a bit of music theory know-how and a strong gut feeling for what works are indispensable.

Here are a few tried-and-true ideas to give your bridge section a jump-start:

- *The standard jazz bridge consists of a cycle of chords that leads back to the tonic of the song.* A chain of ii-V progressions is the typical move. For example, in the key of C, a ii-V is D minor (G7). You could then add another ii-V leading to the D minor, which would make the whole progression E minor (A7) D minor (G7). This is also referred to as "back-cycling," and is such a standard move that you need to know it! To create a useful bridge progression, you could either "back-cycle" far away from the tonic, or simply repeat a shorter part of the cycle. The great part about this is that you can start the bridge with a leap to a far-away chord, but still cycle back in a logical-sounding way by the end of the bridge.

- *Similar to the "back-cycle" of jazz is a chord progression with a chromatically descending bassline.* Again, taking C major as the example home key, the bridge could begin on the IV chord (F). By moving down to E, E♭, D, and then D♭ in the bass, there is a strong pull to the tonic. The tricky part is coming up with harmonies that don't just slip down step by step, which usually sounds amateurish (E major to E♭ maj to D maj, and so on). In a jazz/pop context, an eight-bar bridge using this idea could go like this: F major to G minor to C7 to F major to E minor to A7♭5 (E♭ in bass) to D minor to G7♭5 (D♭ in bass).

In a rock/pop context, those jazzy harmonies may be too much. By adjusting the bassline, you can use a progression that always contains half-step descents, but simplifies the harmonies. A typical rock move is G major to D major to F major to F minor to C. In this case, the bass could move from G to F# (under D maj) to F to A♭ (under F minor) to G (under C). There are loads of these types of progressions that make for great bridges.

Using a *pedal tone* (the same bass note under every chord) is another possibility. The similarity of the note provides a natural harmonic glue to the section, and as other instruments change chords, interesting harmonies occur. Try the verse or chorus progression over a pedal tone bass. Consider playing the progression backward—an older songwriter's trick.

- *If, as mentioned earlier, you're working in a more traditional form (folk, blues, and so on) consider how the different parts can change when the bridge hits.* Background vocals are an obvious bridge addition, and any player can easily add "Oohs and Ahhs" in the bridge. But consider that acoustic guitars can be thumped like percussion, fiddles can play pizzicato lines, singers can play harmonica, anyone can lay out and grab a shaker or tambourine, and—naturally—keyboard players on a MIDI instrument can switch sounds easily.

■ *A great way to introduce a bridge is to drop all the parts that were playing in the previous section, and allow a new part to play solo for a bar, and then bring the gang back in.* Many crafty arrangers use this trick; it's a pleasant surprise for the listeners and immediately reestablishes the connection to the song when the other parts reenter. This concept is as varied as the parts you can come up with to spice up that "break" bar—any melodic instrument, loops, fills, solo breaks, noise, and any combination you can dream up.

■ *If the bridge section starts to sound too disconnected from the song, consider stealing from yourself by adapting a part from the verse or chorus to fit the bridge.* Some writers swear by the "turn it backward" trick for melodies as well as chord progressions. An inverted chorus or verse melody can provide a great connection from the bridge to the rest of the song.

General Arrangement Thoughts

The best way to work out arrangements is by demo-ing the whole song. In many cases, performances can then be perfected, and the demo becomes a keeper. If you plan on working with live drums, you can sequence drums to a click and have the drummer overdub. Even if that doesn't work out—sometimes the energy isn't there when playing to a click—you'll at least have a clear plan.

When experimenting with parts, try tracking each part throughout the song. Then mute them all, and audition them one by one, coming up with an arrangement through the song. If a part becomes boring after a while, see if it can enter later in the song. Consider cutting it in the bridge, changing it in the chorus, or just skipping it altogether. Sometimes after playing around for a while—which is great for learning—it may turn out that the song sounds best stripped down to the basics.

One of the great pleasures of the home studio is the flexibility it allows, especially when working on a DAW/sequencer. Arranging song sections and parts is a lot of fun and, next to writing songs, one of the most satisfying experiences in working on music. When you have the time, experiment freely, reach for those ideas, and play them for friends and colleagues—you'll learn and improve quickly as you get feedback.

11 Mixing: Balancing Art and Craft

Mixing is a craft and an art. Artist's developments in music styles and trends influence their contemporaries—and if the work has true staying power— artists of the future. In this way, one mixer's art becomes the next mixer's craft.

Although this analogy applies to all aspects of music production, it works particularly well with mixing, with the inherent technical side balancing the artistic aspects. Just as a carpenter needs the proper tools, training, and a certain knack for working with wood, there are some basic "must haves" for a good mix—from the tools to the "feel" required to work with music.

In this respect, this chapter is split into two sections. First, the chapter covers the basics required for a good mix. This includes the basic concepts you need to understand and the aspects of mixing that are more or less universal to the process, regardless of the musical style.

Second, the chapter considers the elements of a mix from an "artistic" perspective. Putting standard mix concepts in question can help you find a unique angle with which to approach mixing the music. As more and more people create and publish music, just one new idea can grab a listener's attention and win you a new fan. This section cannot contain any hard and fast rules (for one, they would no longer be "new" as soon as this is printed). As with Chapter 9, "Perspectives on Music Production," this chapter takes a more philosophical approach to help you generate your own ideas.

The Craft of Mixing: Workspace and Monitoring

Although the focus is on the music, you cannot ignore the human element of the mixing process. You will require more concentration when mixing than during any other part of the music production process; mixing is a serious multi-tasking job. You need to be sure that your body is comfortable and that your ears are served well by proper monitoring.

It is very important to note that long-term exposure to high sound-pressure levels can cause hearing damage. *Please* protect your most valuable recording tool—your ears—by monitoring at moderate levels. An average level of 85dB (C weighted) is considered optimal by most professional engineers, and also provides a realistic perceived balance of frequencies to our ears, which do not respond in a linear fashion to all frequencies. Surprisingly, monitoring at higher levels makes bass frequencies *seem* louder, causing you not to mix in enough bass, thereby resulting in *wimpy* mixes! An SPL meter is a good investment in both your health and the quality of your mixes.

Workspace

The most often overlooked aspect when mixing is the workspace itself. Chapter 4, "Setting Up Your Studio," covered room acoustics, and practical, inexpensive solutions for the home studio. Well, if you haven't made yourself some of the life-savers of the audio world—sound absorbers—and put them into place, please flip back to Chapter 4, read up, put down this book, and get to it. Your mixing room needs help.

In fact, if you're ready to mix and still haven't dealt with acoustics, you may be in big trouble! Tracks recorded in a poor acoustic environment are going to be hard to mix. Undesirable room acoustics, room resonances building up on every track, poor isolation, and inaccurate monitoring can make for very poor basic tracks. Mixing is difficult enough—you don't want to be doing damage control at the same time.

In addition to the oft-repeated warnings about room acoustics, you should also prepare to mix by:

- *Setting up at least three monitor systems in your studio.* You'll obviously need a pair of near-field monitors, and there are plenty of good quality, inexpensive monitors available. Beyond that, a pair of headphones and a consumer stereo will be very useful for checking your mixes. Headphones can help you check low bass frequencies if you don't have a sub-woofer or just can't afford to wake the neighborhood to check the kick drum! A consumer stereo or boom box lets you check how a regular, real-world playback system reacts to your mix. This will usually tell you pretty fast if the mix is too dull, too bright, over-compressed, and so on.

- *Making the studio comfortable.* Put the monitors at a good height and angle them so that you are in the sweet spot while working, without straining your neck and shoulders. The "sweet spot" for near-field monitors is where their lines of projection cross (imagine straight lines coming out from the cones and tweeters) and make a perfect triangle. You should sit with the crossing point just behind your head; this puts your ears on axis with the speakers, providing the best stereo image (see Figure 11.1). Be sure that you leave a few feet of space from the backs of the speakers to the walls (and especially corners) to avoid bass build-up.

Figure 11.1 Set up your monitors so that you are comfortably sitting in the "sweet spot" while you work. If you have to lean in or back up your chair to hear properly, that means you're making mix adjustments while not hearing properly—what a waste of effort!

■ *If possible, set up your alternate monitors to work at the flip of a switch.* For example, many home stereos have an aux in on the back panel, which can be wired to a couple of mixer or DAW outputs. You'll use the alternate monitors more often if you don't need to burn a mix and put it on another medium to hear it. Checking on other monitors will seriously improve your mixes. World-class mix engineers all have their favorite "lousy" speakers to check mixes on—whether Auratones, Radio Shack bookshelf speakers, or a Sony boom box—and swear that when they get the mix to sound good on those speakers it translates to the real world.

■ *Organize the physical and virtual studio.* Clean up your outboard gear and connections—patchbays, cables, interfaces, and processors—make sure that everything is working right and there are no buzzing, crackling connections that will sneak into your mix. Keep a pad and paper handy, and perhaps you should take out all the dirty coffee cups and half-empty bags of stale chips as well. On the virtual side, be sure you've backed up your data to a second drive. Organize your files, and make sure you have enough drive space to save several mixes of each song. I also like to grab a few new plug-in demos at this point; it gives me a few new sounds to get inspired to, and sometimes they really make the mix. If the demo is about to run out, consider buying it, or else process the audio to a file so you don't lose that great effect!

Near-Field Monitors

In the big studios there are always a few sets of monitors, including some very large and very small ones. All those professional engineers know this is essential to creating mixes that translate well to the real world. Large monitors provide full-range playback (20Hz to 20kHz), whereas smaller monitors represent different aspects of consumer playback systems.

Contrary to what many novice engineers believe, proper monitors are not meant to sound subjectively good, but should be accurate, and that means *balanced* and *revealing*.

"Huh? But if the mix is good, shouldn't it sound good on the monitors?"

Yes—once you get there, the mix should sound good anywhere—the point is, getting to the stage where the mix sounds good is the hard part. If the monitors sound flattering, as is the case with most consumer products, the mix may sound good before it is actually good. Revealing monitors reveal both the good and the bad. Yamaha NS-10s (although now out of production) are studio standards, and not for their good sound, but rather because they sound harsh in the midrange. If you get your mix to sound good on those monitors, the mix translates well to other systems.

Speakers with lots of low-end thump and clear, sparkly highs sound good, but may be hiding other problems. When you work on speakers that don't flatter the music, you must work harder to balance mix elements, create a sense of excitement, dynamics, and energy, and create a pleasant EQ balance in the lows, mids, and highs. By checking often on consumer speakers, you'll hear how your mix is translating to these types of systems.

In big studios, there is usually a set of large, full-range monitors as well as the console monitors, usually referred to as *near-field monitors*. (Note: Near-field is actually a trademark of E.M. Long Associates. The term has become so widely used that it is, like Band-Aid or Xerox, used in a general way.) For the home studio, near-field monitors tend to be the only practical solution for monitoring due to room size and acoustic considerations. In a nutshell, the idea behind near-fields is to stick your ears right into the sound, thereby reducing the influence of the room enough to hear what's going on.

So, near-field monitors are also a compromise, but so is any playback system, so you learn to deal with it. There is no perfect monitor for your studio, and so I will make no specific recommendations as to what you should purchase. There are fine quality near-field monitors available from $300 to $2,000 or more a pair. Check out reviews, and listen before you buy. I prefer monitors that are very detailed, even a tad harsh in the upper mids, because this makes the musician work hard to be sure the song comes across sounding pleasant. Yamaha's HS monitors have this quality. Other manufacturers to check out include Mackie, KRK, Adam, Yorkville, Alesis, Genelec, JBL, and Behringer.

Look up reviews on the web, and get to a showroom where you can listen, bringing along some of your favorite CDs.

Food for Thought: When checking out monitors, it is tempting to bring along your favorite disks, since you know them so well. Problem is, you already love the music! Music you are unfamiliar with, or test mixes you have made with flaws you know of, make great test material.

You should prepare a test CD with some bass-heavy music, some acoustic music, and some mixes you have created where you know that the high end sounds too harsh, or there was is a "muddy" midrange EQ problem. Additionally, a recording of percussive, acoustic music is a great test for how well monitors reproduce transient details, and a mellow song with a prominent lead vocal should be in there too.

Some brands of monitor aim to impress quickly with strong bass and a sparkly high end. Don't fall for these kind of "hyped" sounding monitors; be critical and listen to how well the monitors reproduce details in the midrange. That hyped sound is fine for home stereos, but may hide flaws in your mix.

By listening for the good as well as the bad, you'll be able to make an informed decision about this critical link in your music chain. There are currently monitors available (the JBL 4300 series, for example) with room EQ correction programs built into the monitors. Although corrective room EQ can add the final touch to your monitoring system, it is not an all-in-one solution. You can't just plop these monitors down in an untreated room and expect the room EQ to save you. Although the response may be flattened out by the EQ, this can't correct for the time aspect of resonances—the offending frequencies ringing in the room will just be turned down somewhat, the masking effect of this resonance will still be there. In my humble opinion, spare the extra few hundred bucks worth of gadgets, and acoustically treat your studio!

When choosing monitors, consider whether you prefer powered monitors, or if you want to set up a separate power amp. Powered monitors may not always have the best internal power amps, which can lead to distortion. Check reviews! The added size and weight of internally powered monitors may also be an issue, depending on your setup.

Small monitors can often have good mid- and high-frequency reproduction, but will lack bass frequencies. Check out the specifications, and find out where the bass end rolls off. You may want to add a sub-woofer to your monitoring system in order to be able to hear what's going on in the 20Hz–60Hz area of your mix. Most near-field monitors don't have a great frequency response in the lowest octaves of the audio spectrum, and you need to be able to judge what's going on down there somehow. Again, in a pinch you can use a decent pair of headphones.

Once you choose monitors, the most important thing to do is to learn your monitors. Spend lots time listening to music in the style you will be mixing on those monitors.

If you have a project coming into your studio in a style you don't usually listen to, ask the artists what their influences are and invest in a few CDs.

The Craft of Mixing: First Steps

As you mix, not only will you be listening intently, but you are also going to have to manage a whole lot of audio tracks, plug-ins, outboard gear, auxiliary sends and returns, mix parameters, edits, crossfades, and all their names and levels. This is a serious task, and anything that can be done to make this easier is welcome.

Start off by cleaning up your audio. Use volume automation and/or trim the audio regions. I prefer volume automation as long as I have enough CPU power. If you edit with a trim tool, be sure to leave enough lead-in and lead-out time on the audio to allow all the natural dynamics of the sound through. For example, cutting off a tom fill too short can ruin the natural decay of the sound. With vocals, the singer breathing in before the first notes is also part of the performance—leave it in there unless it sounds unnatural. See Figure 11.2.

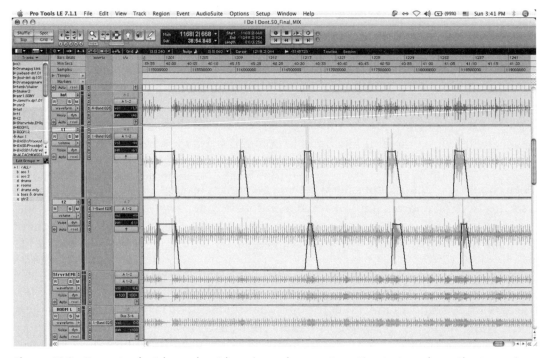

Figure 11.2 Drum tracks "cleaned up" by using volume automation to turn down the tom mics between fills. As simple as this is, it is a key step in making your mix as solid as possible; every bit of stray, out-of-phase audio robs your mix of clarity and "punch." Leave enough space for natural sustain and decay of the sound.

Next, arrange and group your tracks. If you plan on multi-processing any sounds, and then create duplicate tracks, you can always delete them later if you don't need them. Group drum tracks, background vocals, strings, and any tracks that will likely be adjusted together (this is called *subgrouping* in some DAWs; *grouping* in others). Learn the keyboard shortcut to adjust one track in the group without affecting others, as well as the shortcut to "suspend" the group while making adjustments to just one track in the group. Determine a standard plan for your mixer layout. This is a personal choice, but I like to lay out the drums, starting with kick and snare, then bass, the lead vocal(s), harmonies/background vox, guitars, keys, percussion, and then effects. Sticking to a standard plan helps you know just where to find tracks so you can make adjustments without losing the flow.

Plan on reverb and delay sends and returns. When the tracks are organized, set up aux sends and returns. Even if you've printed a lot of effects when tracking, you'll surely be using some reverb and delay effects. It's great to have the sends and returns ready to go as soon as you need them—you can just open a plug in on the aux return, crank up the send on the track of your choice, and you have the effect immediately.

Label tracks clearly. Somehow, I always have one or two tracks of odd overdubs that I failed to label when tracking. This can get really annoying when mixing; these faders stare at me saying, "Remember me? Are you sure?! In which section do I start making noise?" If they are special little parts for the bridge, label them appropriately, such as "bird farts—bridge," so you know just what to expect.

The Craft of Mixing: Building Your Mix

Okay—time to roll the music. Getting started with your mix is a daunting task. Sometimes a mix starts to come together during the overdub phase. If this happens, you're lucky, and you may want to continue working from that angle. Run a rough mix and listen to it in the car and on a few stereos. If you feel you're missing something, it's time to approach the task of mixing with a solid plan.

By the time you're ready to mix, you'll have a whole lot of tracks to deal with. Just listening to all of the tracks together, and tweaking the kick, then a guitar, adjusting the bass, then adding reverb—all in no particular order—may not result in a coherent, cohesive-sounding mix.

Another typical mistake is to solo each instrument, one at a time, and tweak them to sound great alone. When you put them all into the track together, you wonder why the track sounds like mush.

What you need is a plan that focuses on important mix elements. You need to focus on the strengths of the track, and get everyone else into place supporting the plan.

The first thing to consider is the foundation of the track. Is it a jazz/pop piano and vocal piece? A four-on-the-floor rocker? An acoustic folk song? An electro-influenced ambient composition?

Each of these will have a somewhat different focus. Consider each style for a moment. What is most likely the driving element in each? That will be the best place to start building your mix.

For example, in the jazz/pop piano and vocal piece, even if there are drums, guitars, percussion, and strings, these supporting elements must take a back seat to the piano and vocal tracks. Start off by adjusting the piano tone. Listen to your basic tracks, and then determine how to pan the piano mics, if the piano needs any EQ to shape the tone, and perhaps which reverb fits the track. Next, bring up the vocal and find a good balance between the piano and vocal. You've created your basic vibe.

These then become your reference tracks, and other elements in the mix are adjusted to fit to them, not the other way around. You need to keep the main elements sounding natural and solid (unless you're going for a particular effect). If you start to EQ the piano and vocal to make them fit to the drums and strings, you'll adversely affect the tone of those main tracks. It's good to determine these things early on to keep your mixing work focused.

Let's say that you're working on the mix for that four-on-the-floor rock track. Here's a useful standard plan for proceeding:

- *The beat is going to be of utmost importance.* Start by tweaking the kick drum and bass guitar to work together and create a solid low-end. Try to find a complementary EQ concept; by reducing some 400Hz in the kick drum, the bass guitar (which has a lot of definition in that range) will start to become clear.

- *Next, determine the lowest fundamental frequency of the bass, and then set a high-pass filter to cut below that.* Let's say the bass uses the low E string as a pedal note. That makes the lowest bass frequency around 40Hz. By cutting the lows below 40Hz, you make sure that the kick drum owns that territory, creating the low-end "punch" you need.

- *Because the first harmonic of 40Hz is 80Hz, and the next is 160Hz, those are the places to fatten up the bass guitar if needed.* You could boost the kick drum at 60Hz and 120Hz (if it needs more low "thump") without stepping on the bass guitar.

- *Set up a compressor on the kick drum, using a slow attack time.* This will let the attack of the kick through, giving you even more punch. Try to get a lot of attack to start with—as other instruments are added, you'll really need that attack for clarity.

- *Bring up the snare track, and see how it fits into the picture.* Consider its low-frequency content, and be sure it fits with the bass and kick drum. If it sounds weak, consider either boosting lows on the snare, or backing off on your low-end boost on the bass and kick. You don't want to start a "bass race" with every track. Listen to the relationship of the snare and bass in the midrange frequencies. The snare should have plenty of body and "snap" in there. Many modern snare drums can benefit from a wide midrange boost of just a dB or two; they are sometimes built to have a lot of bass and treble, which sounds impressive live, but may need help in the mids when mixing.

- *Adding the vocal will best tell you if you've gone overboard and boosted too much lows on bass and drums.* If you have, it will sound overwhelmed. Boosting a lot of bass on the vocal sounds unnatural, so you'll have to back off on the drums and bass. Consider the vocal level. Try to lay it in the track so that you can understand all the words. If the dynamic varies too much (that is, it's too soft in the verse, and too loud in the chorus), set a compressor to gently correct this.

- *Bring in the guitars!* Luckily, guitars—especially electric guitars—take very well to EQ adjustments. Start by setting the guitars to a healthy volume. Just put them up where they feel good. Then listen to all the other elements. Are they eating up the snare drum? If so, consider dipping out a little bit of mids at the same range you boosted the snare. The same goes for the bass guitar and kick drum; guitars need low end, but they won't drive the song like bass and drums, so consider a little dip in the guitars at important bass and kick frequencies. A high-pass filter can actually make guitars sound "chunkier" as they remove low-end muddiness; so give that a try as well.

This method of determining key elements of the track, and then building the mix around them will rarely lead you astray. In Chapter 11 on the CD-ROM, there is a step-by-step explanation, with audio clips of each phase, from a track called "It Ain't Me" from my band's last album project.

The basic tracks were cut in the "Cash Cabin Studio," the good-sounding but relatively humble home studio of the late, great Johnny Cash. The studio isn't loaded with the most expensive mics and outboard gear in the world, so the basic tracks were solid but not entirely unlike what could be achieved in a home studio. All the overdubs were done using my portable Pro Tools LE/MacBook Pro setup in various home spaces in Connecticut and in Hamburg, Germany. In these cases, compromises had to be made, but with attention to detail and the clever use of some nice acoustic spaces, we managed to get what we needed. Have a listen, and I hope you learn from the mix, and dig it too.

The Craft of Mixing: Setting the Soundstage

The standard for listening to music is still stereo, even though surround sound is becoming more popular. Since people just have two ears, it's best to start by focusing on being able to create a great stereo mix.

By using various stereo techniques, you can to create a sense of the "soundstage"—where each instrument sits within the stereo field—in your mixes. There are a few ways to accomplish this, each way sounds different, and by using different techniques within the same mix you can create a much more convincing stereo placement of sounds.

Obviously, stereo mic'ing techniques have built-in depth-of-field. For example, a "spaced pair" allows you to naturally set the "pan" position of a source; by moving it closer to one of the microphones, the sound playback will "pan" closer to the speaker in which the closer microphone is placed. This occurs due to a complex interaction of volume, frequency content, and time-delay differences. Interestingly, you can simulate these interactions with panning, EQ, and delays—three very simple effects available in even the most basic DAWs.

Let's start with the basic concept of a mono track appearing in the center of the stereo field. This happens when there is equal volume in both the left and right speakers.

The most basic way to place a sound source elsewhere in the stereo field is to use the pan knob. All that panning does is adjust the relative volume between the two speakers. If you move the pan knob 100% to the left, the signal is sent 100% to the left speaker, and the sound seems to move to the left. Although this is easy and mostly quite effective, it is not the only way to place sounds within the soundstage. In fact, it is even possible to make sounds appear to come from outside the speakers, as well as appear as "holograms" in a sort of simulated space between the speakers.

You may have noticed this in some mixes, and wondered what that "three-dimensional" sound is. These stereo placement effects are achieved by manipulating both EQ and time (via delays). Try the following experiments, and you will quickly learn how to create a more convincing sense of soundstage with a variety of effects (you can listen to audio examples in Chapter 11 on the CD-ROM):

■ *Create a duplicate of a mono track.* Pan one 100% left, the other 100% right, and at the same volume. Now, reduce the right channel volume by 6dB. You will hear the image of the sound move to the left. This is how volume affects sound placement, and is the same thing as using the pan knob, and shows how volume differences affect stereo placement. Very simple.

- *To demonstrate how EQ can affect sound placement, put the two tracks back to even volume in the left and right speakers.* Now, create an EQ plug-in on the right channel. Grab the high shelving EQ, and reduce it by 6dB. Again, you'll hear the image move toward the left speaker. Why? Well, it's a bit of psychology, really. Sounds that are farther away tend to lose high-frequency content as the sound wave moves through the air, curtains, leaves on trees, and so on. Clear, high frequencies seem closer. Interesting is that this skews the stereo image in a slightly different manner. The "dulled down" left channel seems farther away; a slightly different effect than volume panning alone. But it gets even better...

- *Here's how delays affect sound placement.* Reset your two tracks, again, to the same level—one 100% left, the other 100% right. Now, create a delay plug-in on the right channel. Set the delay to 10ms (milliseconds), 100% wet, with no repeats. All this has done is delay the start of the entire right channel signal by 10ms. Now play back the tracks. Whoa! What happened? The image again shifted to the left speaker! This *precedence effect* is also based on the psychology of acoustics—if a sound arrives in one ear slightly before the other, we hear the sound as coming from the side of the earlier arrival. This effect works up to about 40ms, after which we start to hear a distinctive echo, and the panning effect disappears. Experiment with different times between 5 and 40ms, and get a feel for how this affects the position of the sound.

- *Now, you can try a combination of all these effects.* This could be called true panning, as it is what's actually happening when we hear a sound placement in the real world. If you're looking straight ahead, and hear a guitar amp off to your left, you are perceiving several "effects" that make that sound appear to be on the left.

  1. The sound is slightly louder in your left ear, due to the proximity of the sound source.

  2. The sound is a bit brighter in your left ear, since the physical mass of your own head blocks some of the high frequencies.

  3. The sound arrives a few milliseconds earlier at your left ear, again, due to proximity.

So here's what you should try. Starting with those "100% left/100% right" mono tracks again, make the sound move to the left speaker. Reduce the right volume by 3 or 4dB. Then reduce the high shelving EQ by a couple of dB.

Tip: Percussive tracks may sound better with more highs cut, to avoid a "pinging" sound, and bass-heavy instruments will sound more panned by cutting lows on the EQed channel. Finally, delay the right channel by 20–30ms. Adjust to taste.

To prove that this is a much more convincing way to set up your soundstage, compare an instrument panned with this "psycho-acoustic" method to a mono track simply panned to the left. This method is not only more convincing, it is that professional sound that you've probably been looking for!

An additional advantage of these methods over basic volume-based panning is that you don't remove the signal from one side of the stereo picture, as happens with normal panning when you pan a sound all the way to one side. With psycho-acoustic stereo, your tracks retain more body and presence in the entire stereo field. Contrasting this type of stereo effect with good old "hard panning" expands your palette of mix tricks, giving your mixes a more complex "big studio" sound.

Phase-based stereo effects are one more way to manipulate the placement of sounds, enabling you to make them appear to originate somewhere outside the speakers. There are dedicated plug-ins that do just this, often called "stereo enhancers" or "wideners." As with all effects, it's great if you know how they really work "inside" the plug-in. This allows you to create the same effect without buying an expensive plug-in that you can easily emulate yourself. It also allows you to experiment with different ways of creating the effect by using more basic effects, which often leads to interesting discoveries.

Stereo image widening plug-ins combine phase-inverted copies of the signal in various ways to manipulate the audio. Try this on a stereo audio file:

1.   Duplicate the stereo file and split the duplicate into two mono audio files, one for the left side, and one for the right.

2.   Process the new left and right split files to invert the phase. Turn their volume all the way down.

3.   Pan the left file to the right, and the right file to the left. Group the files together so you can adjust their volume simultaneously.

4.   While listening to the original stereo file, slowly raise the level phase-flipped, reversed panned pair of tracks. A stereo-widening effect will become audible, making it seem as if the audio is emanating from beyond the physical location of the speakers (farther to the left and right).

The widening effect unfortunately causes center-oriented audio to lose some level, whereas side channel info becomes stronger. Use the effect sparingly on entire mixes, or you can throw the mix out of balance. (This is actually a similar technique as used to remove vocals for karaoke tracks.)

On the CD-ROM, I chose an orchestral string sample (from Apple's Loops) which I found sounds *better* as some of the center channel sound is blended out! Due to the widening effect cutting some of the center channel info, some harsh-sounding aspects of the sample are faded down.

If you find that you are losing certain frequencies when creating this widening effect, there is one last trick you can use to help you maintain frequencies of your choice. Try this:

1. Using the same setup as before, (a stereo signal and its mono splits, phase-inverted and panned to opposite sides), set an EQ plug-in on each of the mono tracks.

2. For monitoring purposes, pan the mono signals back to their original sides (left back to left, right back to right). When you raise the level on the mono channels now, the signal will fade out completely when you get to 0dB. This is because you're adding a signal and its polar opposite together in each speaker. The result is zero!

3. Here's the trick. Engage the EQ on the left channel, and cut 3dB at 120Hz. You'll hear that frequency band pop out at you. Why is that? Because you reduced the cancellation at that frequency! By EQing out the negative, you add the positive! It's like a double negative, a bit confusing.

4. What this can do for you is let you take certain frequencies out of the cancellation equation. Let's say that that string sample started sounding "thin" when you widened the stereo spread, but it was otherwise sitting nicely in the mix. Pan the widening channels as described, set the EQ, and adjust until you hear those frequencies you want to keep. Do this in both mono channels.

5. Finally, pan the widening channels back L>R and R>L, and adjust their levels for the amount of stereo widening you want. You can adjust the EQ to allow more of the desirable frequencies to stay in the center. In this way, you can have the best of both worlds.

Now that's some very fine adjustment! It can come in handy, however, especially for helping parts like strings, synths, ambient sounds, and other odd noises find a niche in your mix. This adjustment also adds to that "wow factor" in your mixes, giving them yet another edge to help grab listener's attention.

Designing the mix soundstage is a unique process from song to song. You may want to sketch out a picture, determining placements from left to right. Then use a combination of panning, EQ, and delay to place the instruments into the stereo field in a convincing way.

In any case, don't overdo the panning and stereo tricks—sometimes a cool part should stand on its own merit. Too many tricks dilute the music.

The Craft of Mixing: Depth and Space

As you adjust mix levels, EQ, and compression for each track, creating the big picture, the mix will start to need a sense of depth and space. If you've planned ahead, you've tracked some ambient mics to complement close mics.

It's pretty common, especially in home studio settings, to close-mic many sound sources. As a result, the mix sounds somewhat one-dimensional and in-your-face, as if all the music lives right up close to the edge of the speakers. This means it's time to put your vast knowledge of effects to use.

Although reverb and delay (and related effects) can provide a good approximation of ambience, it is often difficult to gauge how much to add to an individual track, and how much into the mix.

The trick with effects levels on individual tracks (like chorus on a keyboard, or echo on a guitar), is to highlight without overwhelming. If you find yourself muddying up your mix with too much of your chosen effects, solo the instrument, and find a level for the effect (chorus, delay, reverb, and so on) that highlights the track, but does not mask it. When the level sound balanced, take it out of solo mode, and listen in context. The effect will not seem as prominent. Now don't increase the effect level! If you have placed effects on many tracks, this is especially important.

You may also find that, once mixed and played back on other systems, your mix sounds drier than it had in your studio monitors. This is a side effect of near-field monitoring; near-fields tend to emphasize stereo spread and ambience, so you may become conservative in your use of reverb and panning effects. Try running two mixes, one where the ambience levels sound proper on the studio monitors, and a second one with the reverb returns turned up 3dB. Compare these on other playback systems.

When working with the "spaciousness" of stereo tracks (or stereo pairs of mono tracks, of course), consider the many steps between panned hard left and right, and collapsed to mono:

- A piano or keyboard sound may sound too extreme—to the point of distraction— when panned 100% left and 100% right. Consider panning just 50% to each side or somewhere in between. This can work wonders on synth pads as well. If a pad is standing out in the mix, even as you reduce its level, try collapsing its stereo spread.

In this way, you can keep the arrangement-supporting tone in the mix without the synth dominating the scene.

- The instruments panned farthest left and right always call more attention to themselves. If you're locked into a stereo file, and can't alter the panning, you can use the "split to mono" function in your DAW. Once split, you can treat the tracks as two mono files and pan them as you like. Cubase and Nuendo, for instance, offer independent panners for each channel of a stereo track.

- Sometimes a sound isn't stereo enough! I've often recorded acoustic guitar with two mics, panned them hard left and right, and this somehow still wasn't wide enough. This is the time to open up the stereo widening plug-in, or use the "L/R copy and swap with phase flip" stereo widening trick. This can nicely open up the middle of the stereo field for vocals, bass, and drums.

- To steal tricks from the big studios, spend some time listening to commercial mixes with just one channel on. I find that intros are a great place to learn about how a song is mixed. Listen to the left only, and then the right only. If an instrument is panned left, does it show up in the right channel? Where do the effects appear? When other elements enter, in which channels do they enter? Take notes and emulate what you hear—you will learn a lot in this manner of listening.

The Craft of Mixing: The Overall Balance

Although the mastering phase will make important adjustments to the overall spectral balance of EQ, you should keep an eye (an ear, rather) on the overall balance of your mix's EQ. If you're pumping in too much low end, for instance, you're getting a skewed impression of the mix, and may be dissatisfied with the sound once it is mastered and the EQ balance is adjusted.

A handy tool for comparing your work to pro mixes is a *spectrum analyzer*. There are loads of spectral analyzer plug-ins and programs that display a graph of frequency content and level, either as a snapshot or in real time. I like to select about 30 seconds of my mix, and then 30 seconds of a "pro" mix in a similar style and with similar instrumentation. By comparing the two, you get a visual readout of what the EQ energy in the mix is doing.

For example, if you see a huge bump in low-mid frequencies in your mix as compared to the "pro" mix, you may want to listen carefully, and see if you can remove a bit of low mids from some instruments. This will bring you closer to your goal. Keep in mind, you are comparing your mix to a professional master, which has already been EQ balanced. On the other hand, good mixes don't need radical mastering EQ, so you should be aiming pretty close to that curve.

Food for Thought: Comparing Your Mixes to the Pros   Many CDs made in the last few years have suffered from over-compression. Check out this list of CDs (in all styles from smooth jazz to heavy metal) that engineer extraordinaire Bob Katz recommends for comparison at http://www.digido.com/honor-roll.html. The bar is high with these excellent recordings, but emulating the best is the way to achieve greatness.

Proper monitoring will reveal a lot about your mixes. This includes not only a set of good monitor speakers, but also your D/A converters and the connections leading to your monitors. Running the outputs of your DAW into a crusty old mixer before sending the signal to your monitors may color the sound in a less-than-optimal way. The same goes for the questionable practice of strapping a stereo graphic EQ between your stereo outputs and monitors (the poor-man's solution to corrective room EQ); slight differences in left and right channel gain can confuse the stereo image. In this case, checking your system with a professional CD can reveal problems in your monitoring chain. If the stereo image of a professionally mixed and mastered disk sounds skewed, check your connections carefully to see what the problem is.

An especially useful comparison between your mixes and pro recordings are the relative levels of key elements, such as the bass, kick and snare, and vocals. Also pay attention to the level of reverb used, and the way instruments are panned in the soundstage. By emulating the balance that professional engineers use between these parts, you will be well on your way to developing good mix habits.

In a home studio, getting a handle on the lowest of low frequencies is always the most difficult task. Due to the acoustic problems of small rooms and the limitations of small monitors, you can't always make clear judgments about frequencies down to 20Hz.

There are a few tricks you can use to help you get your low bass where you want it. Using a sub-woofer connected to your regular monitors can help, but you may still have problems in the room acoustics, especially with those high-powered, long-wavelength low frequencies.

■ *Sub-woofer:* When using a sub-woofer, you *must* compare your mix to professional recordings in a similar style. Naturally, there will be radically different levels and timbres of low bass on a jazz tune as compared to a hip-hop or electronic song, so don't compare apples and oranges. Make sure you have your crossover points set correctly. That is, if your sub should be handling 80Hz and below, your near-field monitors should be set to cut below 80Hz. If your monitors reach down to 60Hz, and the sub covers 80Hz and below, you'll be hearing far too much 60–80Hz. If you are blessed with a friend who owns a pimped-out Honda Civic, ask this person to check your bass levels for you, too.

■ *Clever listening:* An interesting thing about low frequencies is that they require more power to amplify than mid and high frequencies. What happens when you start to pump a lot of low-frequency energy into an amp that can't create that much output? It distorts! This is something you can hear in a couple different ways. One way is that you may hear it by *not* hearing everything else as well as you should. If the mids and highs of your mix seem to be in good balance, but the mix is not as loud as you want, there may be lots of ghostly bass frequencies robbing your mix of energy. I use a pair of cheap desktop monitors for this test. By cranking them up, I can easily hear them breaking up if there's a problem, because of their cheap power amp. Well-mastered professional mixes get amazingly loud—even on those cheap little M-Audio boxes—before breaking up.

■ *Eyes as ears:* If you are monitoring on speakers that are connected to an amp with more than enough power to drive the speakers, keep an eye on your speaker's woofers. If they are getting a lot of energy down around 20Hz, then you may see them slowly pumping back and forth without your hearing anything. Time for a high-pass filter; these frequencies are mix killers. This can happen on any monitors, although I tend to see it more readily when there's plenty of "clean power" available. Using a spectrum analyzer will also help you see these naughty waves.

Furthermore, keep a perspective on the mid and high frequencies. Every microphone, processor, and plug-in has its own signature sound curve, shaping the tone of whatever sound passes through. This could be a mild treble boost, or a "thickening" of the tone, for example. As you process more and more tracks with that particular device, this sound curve can become prominent in your mix.

This is a phenomenon the big studio engineers are well aware of, and they use it both for positive effect (such as tracking all the basic tracks on a great old Neve console for its particular sound) and consciously avoid it (by using different microphones and outboard preamps during overdubbing). In the home studio, you may be limited to a small selection of gear, but be sure to use all you've got! Just because a boutique compressor or set of new plug-ins you invested in are top quality, doesn't mean you should slap them across every track you have.

When comparing your mixes on other systems, take notes. If the highs are Frito-Lay crispy, or the lows are at elephant-flatulence level, take notes. This is telling you something about your monitoring situation. Whatever you hear too much of, you're not hearing well enough in your control room, and vice versa. An old trick for mixes that sound dull on other systems is to tape a layer of tissue across the control room tweeters. You hear fewer highs, so you boost 'em a bit more. This leads to a brighter mix when listening outside the control room.

Proper room acoustics also reduce listening fatigue, so keep those acoustic treatment projects from Chapter 4 on your to-do list!

As a final reminder for this section—take frequent breaks when you're critically listening. Not only will this save your hearing, but moving around once in a while will keep your neck and shoulders from cramping up, thus maintaining your concentration.

A short break from the music will also keep your perspective on the task at hand. If you've been editing the same percussion part for 45 minutes, and everyone is becoming unsure if it's right, a walk around the block may be exactly what's needed. You must keep your ears fresh. If your start to go numb to high frequencies from ear fatigue, it's as good as a guarantee that your mixes will not translate well to other playback systems.

The Art of Mixing: Overviews, Strategies, and Impressions

The technical side of mixing is like the Olympics of multitasking; parts, levels, EQ, compression, reverb, effects, automation, and outboard gear... You've learned how this can be approached with a plan, making this variety of tasks more manageable.

The technical challenges are just one side of the coin, however. You could think of it as the objective side—you do your best to get a great drum sound, create a sense of depth and space, balance the mix elements, and so on. This may well be 95% of the job, but there is always room to make the mix stand out just 5% more, putting its head above water in the ocean of music out there.

There are many ways a small touch can make a track stand out. One useful way to put the final touches on a track is to brainstorm on paper, and then approach the music with fresh ears. This is also particularly useful when tracking and editing vocals. Large sheets of paper taped to the wall and a black marker are mainstays of mix work in many studios; it helps you organize thoughts and ensures that small important details are not overlooked or forgotten.

For example, a "happy accident" like a wrong aux send turned up—for blast of reverb on the vocal track at an odd point in the song—can bring a unique twist to a track. Note this on the list, and experiment in the mix.

Here's yet another list of ideas to consider. Brainstorm freely, listen to songs you like, pull out ideas, and try them. If something doesn't work, cross it off, reduce possibilities, and distill the mix down into its most concentrated spirit.

■ Once you have your basic balance of all elements, and the mix has reached that point where everything blends, listen at different levels. Turn it way down. Crank it. Now try featuring an important element by turning the level up a few dB. It's surprising how loud the snare drum often is in many big studio professional mixes. Try the same with different tracks (lead vocal, rhythm guitar, and so on), and see if one of these variations makes the mix really come to life. (It will probably be the snare.)

- If you've kept scratch tracks—especially rough vocal takes from tracking—try listening to them again. It often happens that the energy of these tracks adds back an edge that gets polished away over the course of overdubbing and mixing. Try adding just a single line or word from the scratch track into the lead vocal here and there. If you listen carefully to The Beatles' "Revolution" (the "Hey Jude" B-side version with the fuzzed-out guitars), Lennon's vocal is doubled at odd spots in the verses, creating a unique energy that defies definition.

- Since you now know how to create many effects using basic EQ and delay, do just that once in a while. Skip that fancy new plug-in for an old, funky outboard reverb— or set up an echo chamber in your bathroom by using a spare monitor/amp and a microphone. Some older outboard processors offer effects that can't be matched by plug-ins, like the Eventide H3000-d/se, or are just uniquely crappy, like the Boss SE-50.

- Ambience can be created many ways, and can bring a new level of depth to a song even if it's mixed in at subtle levels. If the song, arrangement, and mix are in shape, but there's still something missing, consider adding some ambient effects. Sample the static from a record player or AM radio, and lay it under the whole track. Alternately, blend it with just one part, so it enters and exits with just that part. Record the sounds outside your studio. Process them with pitch shifting, delay, reversing, filters, distortion, and so on, until you find a texture you like, and add this to your mix. More harmonically fitting ambience can be created with synths, of course, but a synth-like ambient effect can be created from any chordal instrument with copious amounts of compression and reverb. You could even use only the effected part of the signal for less obvious effect. The Verve's "Lucky Man" intro makes use of ambient vocal effects in this manner.

- Blasting tracks back through a PA into the tracking room and re-mic'ing this ambience is more widespread in big studio trickery than you'd imagine. This was used to enhance the drum sound on the recording of Bruce Springsteen's "The River" and Soundgarden's "Superunknown," for example. The big advantage of this in home studio situations is that you can set up the effect, and then run the tracks through and record the results when you have the place to yourself. Use the biggest room you can, and if there are wood floors, all the better.

In the philosophy of music production, one of the best bits of advice (drawing from the well of The Beatles yet again) is to "avoid foolish consistency."

If a mix fails to hold your attention, there may be something that is too consistent throughout. Consider consistent elements:

- Is the dynamic the same for too long? Try these tricks in selected sections of the song. Vocal dynamics can be livened-up with simple volume boosts or delay doubling effects. Drum tracks get an energy boost from samples and reverb effects. Doubling the bass with a synth line adds size and density to the low end. Delays that aren't timed precisely to $\frac{1}{4}$ or $\frac{1}{8}$th notes create rhythmic tension. Suddenly removing reverb from a part creates a dynamic change as much as a volume/intensity move.

- Have the drums sounds been sterilized with samples and processing? A track that uses the same snare and kick sample throughout the song will lose the listener's interest by the end of the first chorus. Map out the sections of the song and swap, remove, or change the volume of samples at section changes. Ah! Dynamic change!

- Have you cleaned up the song too much in general? A lot of the life of a song's performance lies in the singer's breaths between lines, finger slide noise on the guitar, and the rough edges of parts entering and leaving (such as the pickup sounds of percussion parts). Try removing some of your "cleanup" cuts and fades. This may just bring the human factor back into the song. DAW editing can easily go too far.

- Songs can be harmonically "too clean" as well. Sticking closely to simple harmonies can be great in a "less is more" way or just boring. This is better addressed during rehearsal and arrangement, but it's not too late even at mix time. By adding a MIDI synth simply sustaining one of the basic chords through an entire section of the song, the basic harmonies changing over top of it may take on a new life, even if this sound is mixed subliminally low. Try it out; use a string, organ, or unobtrusive pad sound, and hang on to the I, IV, or V chord (try anything!) and see what happens.

- This goes for tuning as well. Pitch-correction software (AutoTune and the like) is great for fixing pitch problems on otherwise great performances, but can scrub a performance clean of its beautiful affectations. Consider automating pitch software to leave in the occasional pitch drop after a line or to leave some sections raw and nude. Some singers should never be tuned—imagine piping Mick Jagger or Bob Dylan through AutoTune, and you can imagine the smoldering shell of one unlucky DAW. Beyond vocals, take advantage of the thickening effect of tuning modulation on synths, chorus effects, and pitch shifting duplicates of a track up and down a few cents, and panning them left and right in the soundstage. Eventide's H3000 has been used on countless records for its detune/thickening effects; it's one of the "go to" mixing tricks today, even though the processor has been around for decades.

■ Is the song structure too predictable? One of the great advantages of DAWs is the ability to non-destructively edit entire arrangements. Just be sure to save your work before editing, and save edited arrangements under a new filename, so you'll always be able to get back to the original. If that last verse seems to drag on, try cutting it in half. Too many repeats of the chorus before the solo? Whack—fixed. Need an intro after all? Copy and paste in the chorus, verse, or bridge, and experiment by removing parts and creating sound collages. Sometimes just laying in a single sample of the guitar hook, background vocals from the chorus, or a drum fill taken from later in the song makes a perfect, understated intro.

■ There are numerous simple ways to avoid consistency. Mute the vocal reverb in part of a verse. Remove the double of a guitar track for a few bars. Drop the bass in the first half of the second chorus. Remove all but the snare drum for a few bars. Sometimes that's all it takes.

The flip side of the previous bit of advice is, "don't reinvent the wheel." It's tempting to use all the readily available functions in your DAW to process and manipulate your audio. However, too many unusual sounds and mix moves in one song can be just plain confusing to the listener. Outrageous sounds are a great way to grab attention, just consider presenting them in an accessible way. In practical terms:

■ Most listeners, as noted earlier, don't hear the separate aspects of the mix, but rather the blend of all the elements. If you want to keep the focus on the lyric, be aware of how much attention other parts are getting. Basic, solid sounds will work best in most songs, keeping heads nodding and feet tapping. Sounds that are filtered, flanged, fuzzed, and otherwise affected draw attention. Start off with great basic sounds that support the leading elements. If, after working this way, you feel more fireworks are needed, go ahead and add them. A great example is U2's "She Moves In Mysterious Ways." The Edge's flanged guitar part grabs the spotlight at the top, but makes way for the vocal at the beginning of the verse. It isn't until later in the song that guitar, bass, and vocals all overlap. This provides an aural surprise around every corner without sounding overloaded with tricks.

■ If you opt for a strange snare sound, that will grab the listener's attention. Take a few bars to introduce this sound. If, at the same time, you add filtered and flanged guitar, outside harmonies implied by the bass, and vocals performed through a vacuum cleaner hose; you may be asking too much of your listeners. Consider ways to introduce these elements and get the most bang for your buck. Start with the drums, bring in the guitars and bass, and then drop them all for a half a bar when the vocal hoovers into the mix. Perhaps this is brilliant, revolutionary music. On the other hand, are you just hiding unsatisfactory performances? If your aim is to confound the listener, is your mix truly obscure and astounding? Be self-critical.

■ Clichés can be useful. For example, Oasis's "Champagne Supernova" starts with a collage of guitars and water/ocean effects. Perhaps if the title had been "Ocean Landslide" the water sounds would have crossed the line into kitsch. Oasis has a knack for using clichés; clever use of expected elements is also important. Do you think they went too far with the intro to "Don't Look Back In Anger," compared to Lennon's "Imagine"? Maybe that is clever; the pop musician as clever thief.

■ Play your mix for others at a party, dinner, or in the car. I find that these situations force me to hear the music differently, in the context of communication. This is often the point when it becomes obvious if there are too many fancy production tricks, or if the track needs more spice to keep listeners interested.

On the Way to Mastering: A Mixing Checklist

There are also a few last dos and don'ts when mixing. These points are meant to help you deliver the mastering engineers what they need to do their jobs. Some are obvious; some are not. I've screwed these up so many times that I keep a paper checklist handy when making mixes—these are exactly the points you might forget when you're caught up with making fractions-of-a-decibel level adjustments and tweaking reverb decays for the fourteenth time when mixing!

■ *Check the track for noises and glitches.* Check for vocal sibilance and pops, particularly harsh string squeaks, digital zaps and bleeps, and so on. These are the mixer's responsibility, not the mastering engineer's. Oh, he'll fix them—but at his prices!

■ *Be sure that you've built a strong foundation.* Listen at both loud and soft volumes and be sure that your mix properly features the main elements. This can't be fixed in mastering—now's the time to double-check your mix concept.

■ *Are any unnecessary devices/plug-ins taken out of the mix?* Mute aux inputs when they are not in use, disconnect mics from mixers, and so on. I've made embarrassing mixes with director's commentary, by accidentally leaving a mic connected to the console!

■ *Once in the digital realm, stay digital.* Avoid extra A/D and D/A conversions whenever possible. Additionally, when going digital, use the best converter you can get your hands on. If you've recorded analog and are mixing to digital, consider renting a top-notch converter for mixing; it will drastically improve sound quality over an inexpensive USB audio interface's converters.

- *Is there enough headroom?* With the dynamic range of 24-bit recording, there's no reason to push everything up to 0dBFS. The mastering engineer can make your mix seem as loud as you like. Because your meters may not be completely accurate, leave a few dB headroom over the loudest peaks, with most of the mixes peaks kicking around –6dBFS. This is a model argument for "better safe than sorry." If, at the mastering studio, you hear digital overs on snare hits that you hadn't heard when mixing, you'll be wasting time and money.

- *If you use a stereo bus EQ and/or compressor, run a second mix with these effects removed.* The overall mix EQ and compression are best left to the mastering engineer. You can always recommend the plug-ins/units you thought sounded great on the mix; a true pro mastering engineer will listen. However, ear fatigue when mixing may cause you to start boosting high end and over-compressing, skewing the mix and creating more work for the mastering engineer. These sorts of overall balance decisions should be left for the mastering stage. If the mix sounds lousy without your stereo bus EQ, you need to go back and work on EQ settings track by track! Be sure to clearly label each version you deliver to mastering, so as not to confuse processed with unprocessed versions.

- *Mix at the highest possible resolution.* For example, if you tracked 24-bit 48kHz audio, then mix at 24/48 as well. There's no point in mixing at 24/96 in this case, as it won't increase resolution. Definitely don't reduce to 16 bits or change the sample rate. Because internal processing of the audio is occurring at resolutions even higher than 24 bits, you should add dither even if you are staying at 24-bit resolution when mixing.

- *Don't create fade-ins or fade-outs.* Leave this for the mastering stage. Make notes for the mastering engineer, such as "begin fade at 3:23, fade out until 3:34." Several steps in mastering need to occur before fades are created. If you create a fade-out, for example, this will create a problem with the mastering compression as the overall level fades down. Fades must be created post-processing.

- *Create alternate mixes.* If you're at all unsure about the overall level of, say, the lead vocal or the snare drum, create alternate mixes labeled _vox_up, vox down, _snare_down, and so on. Ask if the mastering engineer shares your opinion about the level, and pick the right mix at that point. It may cost a little extra, but is still easier than a remix. Another industry custom is to run a mix with the vocal muted. This may be useful for ads, television, or in the case of pitching a song to another performer. The mastering engineer will probably be willing to run the "no vox" mix through the same settings as the main mix at minimal cost.

- *Stem mixes.* If you are having a hard time finding balances because of poor room acoustics, or if your monitors aren't up to snuff and you feel unsure, you can create *stem mixes*—separate sub-mixes of the drums, instruments, and vocals—and let another engineer create a balance between those. Be sure to ask if this is a service they offer first, but most will be happy to help. This allows you to add your own panning, effects, automation, and so on, but still have the flexibility to adjust the basic balances in another studio rather easily.

- *Get outside opinions.* Before running off to master the mix, play it for a few trustworthy acquaintances. Just as you've tested your mix on several different sets of speakers, try it out on some different ears. Make sure you're getting the message of the song across. It's your last chance!

- *Crack a cold beer.* It's quitting time.

12 Mastering: The Final Adjustments

Since the dawn of the all-in-one mastering gear units like the T.C. Electronics' Finalizer, the audio recording world has been abuzz with the word "mastering." It seems that everyone had been unaware that there was a step after mixing that somehow improved recordings before distribution, or at least that this was not considered a significant process.

Not surprisingly, mastering has become hyped into the next "holy grail" of the home studio, and the newest industry buzzword. The audio plug-in market now offers dozens of "all-in-one" mastering plug-ins to meet demand. Now we must all become mastering engineers as well as composers, performers, and tracking and mixing engineers!

If you can manage to catch the ear of a professional tracking or mixing engineer, ask this person if they'd "just have a go" at mastering, with their reputation on the line. If you've asked a serious professional, the answer will be a resounding, "No way!"

Just as the process of mixing involves balancing relative track levels, compression, EQ, and effects, mastering requires a different group of listening skills. It's not necessarily a process that only the fabled "golden ears" crowd can perform, but it does require unique skills.

The mastering engineer takes on several tasks, all of which influence the final impression of the music. Each track is optimized in terms of EQ, compression, limiting, stereo spread, mono compatibility, and that elusive ability to "translate well" to as many playback systems as possible. Naturally, a good mixing engineer keeps these things in mind. But doing this doesn't do away with the need for mastering! Proper mixing changes a mastering engineer's job from "how do I rescue this mess!?" into "let's make this well-mixed album sound fantastic."

Beyond that, the mastering engineer adjusts relative levels between the tracks, creates fades, spaces the tracks relative to one another, and provides the proper media format for delivery to the production house. It is a complex process that cannot be done properly "on the side" by just any audio engineer.

At this point, I must state that I am not an experienced mastering engineer, but a musician and producing engineer who likes to try his hand at mastering whenever possible. I'm no authority on this topic, and don't claim to be—this chapter is meant to aim you in the right direction. Hey, everyone has to start somewhere! If mastering interests you, I recommend you read an entire work dedicated to the subject, such as Bob Katz's *Mastering Audio: The Art and The Science.*

Mastering Rules

These rules are "best practice" conventions to use when mastering, and are recommended industry-wide. Ignore them at your own risk!

- *The mix engineer should never be the master engineer.* This is not to say that a mix engineer can never be a mastering engineer, or vice versa. That's certainly possible, just not on the same project! In the same way that different speakers reproduce music differently, a different set of ears will hear the same music differently. Hmm, maybe we can make latex casts of George Massenberg's ears and sell them along with plug-in bundles. I thought of it first! Imagine the ad in *Mix* magazine. In all seriousness, a second engineer with trained hearing helps the overall recording by balancing out quirks and hearing and correcting problems the mix engineer may have missed.

- *The mixing studio should not be the mastering studio.* Ugh... this is getting rough on the home studio engineers! Although disappointing, this is also logical. Every studio has its acoustic imbalances, from the shape of the room to the characteristics of the monitoring system. Part of the mastering process is to even out these imbalances, and this naturally cannot be done in the room with those problems! Imagine that your studio tends to overemphasize frequencies around 250Hz. When you mix, you'll tend to cut these frequencies on all tracks, making the mix sound fine in your studio. If you then try to master the mix in that room, you won't hear the problem, and the mix will not translate. By moving to a mastering studio, the problem becomes apparent, and your mix is saved.

- *Deliver your mix in the format required by the mastering studio.* Give the studio people what they ask for—digital tape, digital files, or analog tape—they know what they need. If you can't deliver digital tape, for example, tell them this and find a solution beforehand.

- *Don't make "loudness" your first priority when mastering.* First of all, experienced mastering engineers will love you for not demanding "loud above all." Stereos have volume knobs for making music louder, so you should focus on dynamics and overall sound quality, and then see if you can make the master louder after this. Face it, someone else will always be louder than you; you can't compete here. So why not make the music sound great instead of just loud?

■ *Don't fall for all those offers for "cheap, fast, loud mastering" that are all over the Internet and in trade magazines.* You can do the same as most of those offers can with those "hot master" plug-in presets. If you have a budget, contact well-known mastering houses for price quotes and estimated delivery times. Some $50 per track "deal" will not get you more than you can do at home. Fifty bucks won't buy the attention of an experienced professional, so save your money.

With these points in mind, you're now ready to *not* master the music you just mixed, right? Assuming that you'd never break the first rule mentioned, let's learn some basic tips on how to master audio recordings. Keep in mind, there is no right or wrong, and two professional mastering houses will deliver two different sounding masters. If you are a determined do-it-yourself engineer, make it your goal to make the music sound as good as it can on many playback systems through the following steps.

A Step-by-Step Introduction to Mastering

Begin by organizing all the audio to be mastered. Be sure that files are in the proper format, and have been dithered (even when mixing to the same bit rate, due to internal processing), are at the proper sample rate, in the proper file format (.wav or .aif) and haven't been converted to MP3s (surprisingly, this happens!). Create separate folders for mixes and the resulting masters. Label your folders and files clearly, and create backups.

Considering that one unit of proper mastering gear probably costs more than your car, I start with a basic assumption: *You have what you have, and that's all that you have.* You're not going to drop $6,000 on the mastering version of a Neve 33609 compressor. Although that's what a top mastering house will use, you can only use the best gear (or plug-ins) you have available. This will be just one compromise of mastering at home!

INSIDE THE BOXES
What's Inside Those Boxes!?   Q: "Pray tell, dear author, why do mastering engineers have one EQ unit that costs as much as my whole studio? Are they really so superb that my recording is crap in comparison?"

A: Dear concerned home studio owner and gear fanatic, as you've learned through-out this home studio recording guide, pretty much everything—gear included—is subjective. That funky old Pultec EQ or dbx compressor you use that sounds so cool on your tracks is good. A mastering quality EQ or compressor probably wouldn't give you the sound you desire when tracking drums and doing overdubs.

This is where one of the major differences between the recording studio and the mastering studio becomes obvious.

In the recording studio, the main concern is to capture performances and create great-sounding tracks in the process. Anything goes, as long as it sounds good. Funky, dirty gear has as much a place there as does a few spankin' new top-of-the-line preamps, EQs, and compressors. You then create a mix, blending all these elements into a harmonious, masterful work of art. Then it's off to the mastering studio for an objective listen.

In the mastering studio, there may be some of the same gear in the racks, but it is often built differently, and will be applied differently. A "mastering version" of a given EQ or compressor uses stepped-gain potentiometers, which add significantly to the cost of the unit. This is necessary for accurate recall of settings in case you need to re-master any of the songs. The unit must also be a meticulously engineered stereo unit (for proper imaging), during tracking where multiple mono units can be used in a stereo linked pair. Again, the cost increases.

Additionally, the "transparency" of the audio path is of paramount importance when mastering. An EQ that sounds "warm and full" on a single track might not be acceptable for use on an entire mix, because it might cause phase shifts and add color that skews the overall sound of the mix. Mastering gear must maintain the impression that the mix engineer worked so hard to create. The goal is to correct EQ balance and adjust the dynamics in an unobtrusive way that improves the impression of the music. In a way, it is the objective side of gear!

Beyond the audio gear toys that recording engineers dig, there are top-notch A/D and D/A converters, digital master clocks, two-track master audio recorders, full-range audio monitors, and expensive acoustic treatment and isolation systems at work. These are pricey toys that you might not find in the recording studio.

To wrap it up, there's no need to be jealous of all that fancy gear; you don't need it! The double and triple cost of mastering gear might improve the performance by only 2% in some cases, which is probably not necessary in your studio.

Next, be sure that your monitoring environment is in the best shape possible. Mastering is about listening and making informed decisions about the mix, which can only be made with reliable information. More reliability means better monitoring, which in turn produces better mastering. Again, this is a compromise, and you need to minimize the compromise by using these guidelines:

- *If you have extra sound-absorbing panels or bass traps in your tracking room, bring 'em in.* You won't be able to create a perfect listening environment, so the rule of thumb "more absorption is better" applies. It's better to catch those naughty waves and absorb them.

- *If you're using near-field type monitors (most likely), you'll need to be able to hear all the way down to 20Hz, so a sub-woofer is necessary.* As you learned before, bass frequencies take more energy to amplify than mids and highs, and take the lion's share of energy on playback. If there are still rumbles and other non-musical bass frequencies in the stereo mix, mastering is the time to remove them. If in doubt, I'd dare say you should cut everything below 30Hz. There's very little musical info down there, so it's a pretty safe bet, and you stand to gain more than you lose.

- *Prepare your visual monitors.* These include the dBFS and dBVU meters as well as the spectrum analyzer and phase plotting plug-ins. There are free, top-notch dBFS metering plug-ins available from www.solid-state-logic.com (AU and VST) and www.smassey.com/plugin.html (RTAS/TDM). You'll need to keep a meticulous eye on level when mastering, and the stock meters on many DAWs aren't very accurate.

- *I noted earlier that I'm not a fan of EQing monitors in home studio situations, but if you've attenuated the problems with acoustic treatments, the additional effort can squeeze a bit more accuracy out of your room for mastering.* You'll need a reference microphone and real-time analyzer/calibration software. Compensate for problems gently; serious EQ problems are often rooted in resonances (time domain), and cannot be fixed with EQ alone (frequency domain adjustments).

- *Let CDs be your guide.* Just as when mixing, comparing professionally mastered mixes to your work is key. Beware of MP3 files for this purpose; they may be compressed, EQed, and have odd-sounding artifacts in the high frequencies. Pick a few tracks that are similar in style and mix content, and import them into your DAW. This way, you can be sure that you are comparing levels accurately. You can also run spectrum analysis on the professional mixes and compare them to yours. For exemplary recordings to use for comparisons, refer to Bob Katz's honor roll at http://www.digido.com/honor-roll.html.

- *Don't neglect your ears!* You should be rested and relaxed, not just rolling in from the encore of the Black Sabbath reunion. Needless to say, use earplugs at concerts to maintain the pleasure your hearing provides you.

Tip: A great way to check your mix for radio-compatibility is to rig up a faux-radio playback. Radio (too often) heavily compresses and limits the program material, so radio play will be a trial by fire for your mixes if you haven't tested them in this format! When you have a working mastered mix going, play it back with heavy multi-band compression and a limiter, reducing peaks by about 6dB. This approximates the atrocities that radio will commit to your music. A typical problem this may reveal is an abundance of bass energy. On capable playback systems, lots of bass sounds cool, but as soon as compression comes into play (from radio or from overloaded power amplifiers), the low end will begin to dominate the music, creating a "pumping" sound correlating to bass instruments. Take this as a hint to reduce low frequencies if this happens in your faux-radio test.

Mastering Processing: Practical Tips

Again, the advice here is meant to be practical for those who want to master their work in a home studio environment. The big studio pros are naturally cringing right now, but so be it. They got their start somehow, somewhere. All but the over-privileged have had to work in less than ideal situations. Put that in your pipe and smoke it, hotshots.

So let's dig in and see (hear!) what can be done with your mixes.

There is logic behind the chain of processing when mastering. Whether setting up physical hardware or software plug-ins, the concept of "gain-staging" is important for optimal audio quality. You should keep all units/plug-ins within their optimum operating range to avoid self-noise (at low levels) and distortion (at high levels).

The typical chain of processors when mastering is Audio > EQ > Stereo Enhancer/ Widener > Compressor > Limiter > Output.

It doesn't make sense to add a processor that may increase gain after the limiter. The limiter essentially puts a cap on the output, so a change in gain after that will either ruin your plan to maximize the level, or push you over and into digital clipping.

That's why EQ generally comes early in the chain. EQ usually precedes compression, as it is best to feed compressors a well EQ-balanced signal. Monstrous bass drum sounds could trigger the compressor into "pinching" the whole mix (which is also a hint that the low end needs some mastering EQ!).

Stereo enhancing/widening effects could come after the compressor, but again, not after the limiter, which is always last.

This doesn't leave a lot of room for experimentation, which is fine. You'll have your hands full trying to decide which adjustments to make on these processors anyway.

Food for Thought: In some cases, EQ is used after compression to counteract the "dulling" effect compression sometimes has on the audio. This, again, is a matter of using your ears to determine the best way to proceed. You may even find that using an EQ before the compressor *and* afterward (to slightly boost the high EQ) is the right combination.

In any case, as you get the knack for listening to these subtle differences in the ordering of mastering processors, make a few alternate masters using these variations. Spend some time listening to them on various playback systems to discover how these fine adjustments sound.

Master Equalization

Equalizing a stereo mix will quickly alter the overall impression of the music. Before you start EQing anyone's mixes, I suggest that you play around by EQing some professional's mastered mixes! That's right—take a favorite song, copy it into your DAW, and experiment with EQ.

What you'll quickly discover is that (other than radical moves) nothing sounds particularly wrong. A good mix can't be hurt much by an adjustment of a few dB in the lows, mids, or highs; but it does sound somewhat skewed. Try boosting and sweeping EQ through the whole range of the EQ spectrum. Find out for yourself which EQ bands sound warm, muddy, nasal, harsh, tinny, and so on. As you hear these qualities in a mix you are mastering, you'll know where to begin adjusting the EQ.

Next, try turning your monitors down to a fairly low listening level. You'll find that boosting both lows and highs sounds correct when the volume is low. With this EQ boost applied, turn up the volume. The bass and treble will now sound out of proportion. This is the principle of "loudness" in EQ. Basically, our ears are more sensitive to mid frequencies, so a bass and treble boost sound proper at low levels. For this reason, it's important to monitor at about 85dBSPL (83dBSPL is theoretically the "perfect" volume although it's pretty hard to nail exactly 83!). An inexpensive dBSPL (sound pressure level) meter from Radio Shack will keep you on target.

There are no rules for mastering EQ such as "always boost 16kHz by 1.75dB." The need for EQ will vary in as much as musical arrangements and the quality of mixes vary. However, the following tried-and-true EQ advice is as pertinent as ever:

- *Choose cutting over boosting, it is more transparent.*

- *Choose gentle EQ slopes (wide bandwidth) over narrow boosts or cuts; they are more gentle sounding.*

- *Cutting below 30Hz will rarely hurt your mix, and may solve inaudible low-frequency problems.*

- *If a particular bass frequency seems to resonate within the mix, a narrow bandwidth cut of a few dB may be in order.* As you've learned, many small studios have particular bass frequency problems. Be sure to triple-check this decision on other monitors to rule out the problem being a part of your monitoring system!

- *The magic "mud" frequency is 360Hz.* Engineers who haven't discovered this may have too much of this low-mid range crowding their mixes; others may become overzealous and cut far too much, leaving a hole in the mix. Consider this critical frequency, and compare yours to professional mixes.

- *Above the "mud" and below the harsh upper mids is the 500–1kHz range, which makes up the body of the mix, and to which our ears are rather sensitive.* A broad boost of as little as 1dB in this range often makes a mix sound more "forward," and helps it translate better to smaller speakers. If the mix sounds nasal, however, this is the range in which to cut back, in particular near 1kHz.

- *Watch out for harsh upper-mid frequencies.* Many inexpensive mics have peaks in the 2–8kHz range, and as these tracks pile up, the mix may become harsh. Cutting in this range can unfortunately interfere with the clarity of vocals. In this case, using a dynamic EQ—that is, a multi-band compressor—may be the best solution. Use EQ to "boost and sweep" and find the offending frequencies. Next, engage a multi-band compression plug-in and adjust the high-mid EQ band to correspond to the naughty frequencies. Start with a gentle compression ratio (2:1) and a very high threshold (so that the compressor is not working yet). While listening, lower the threshold until the compressor just starts "cooling off" the harsh frequencies. Take an ear break for a few minutes. Come back and listen to the mix while turning the multi-band compressor on and off. It should not be cutting more than a few dB, and this should do nicely. Radical cuts in this range are trouble; if the mix requires that much surgery, you'll need to order a remix.

- *High-frequency EQ is a tricky subject.* Although boosting highs gives the impression of clarity, it happens that most professional masters have a gentle high-frequency roll off, not a brash boost. Properly recorded tracks, and therefore well-crafted mixes, have a clarity that goes beyond the "sizzle" of high frequencies. I've heard many mixes with high frequencies that could microwave a burrito in front of the tweeters. After adjusting the highs down to terrestrial level, the mix sounded simply dull, as if there was a blanket over the speakers. What to do!? Well, I've stopped hair-pulling over these situations—"garbage in, garbage out"—poorly recorded tracks are hard to rescue! In this situation I suggest careful application of an exciter and/or multi-band compressor as a dynamic EQ. Set the multi-band compressor to *expand*, not compress, those highs.

- *Up in the "air."* Due to the combined drop in both our hearing's sensitivity and playback system reproduction of frequencies above 15kHz, some engineers like to add a healthy boost up in that "airy" range. Give it a try, and trust your ears. And your dog's ears—if man's best friend hides under the hamper when you crank up 18kHz, reconsider. Be sure that you're not making the mix sound brittle or tinny.

- *To increase the apparent loudness of a particular instrument, try gently boosting the harmonics of that instrument's musical range.* If you hear many balance problems in the mix, however, it will be better for your reputation if you suggest a remix rather than carving up the mix with EQ.

Stereo Spread Adjustment

Operating on principles such as the Haas Effect and M/S decoding, stereo wideners seem to do the impossible, placing audio outside of the speakers. A mild widening effect can certainly make your masters sound more professional, but beware the novelty effect of new toys. What is exciting at first quickly loses the "wow" effect and becomes distracting.

As most stereo widening plug-ins are designed to be mono-compatible, you're generally safe to use them when mastering. Waves' S1 Widener plug-in is a popular choice for mastering applications. It's possible to use the technique learned earlier, wherein you pan a copy of the left channel to the right, the right channel to the left, and flip the phase on these copies. This is based on the M/S decoding concept, and is mono-compatible as well. Just remember that it does begin to reduce centered audio as you increase the effect.

When dithering to 16 bits, there can be a minor loss of stereo depth, and a touch of stereo widening helps maintain the original 24-bit stereo spread. In my opinion, this is just icing on the cake, and won't make or break any mix unless used to excess.

Compression in Mastering

I'll start off this section by quoting Bob Katz's rules:

> "Rule #1: There are no rules. If you want to use a compressor/limiter of any type, shape, and size in your music, then go ahead and use it.
>
> Rule #2: When in doubt, don't use it!"

This joking oversimplification helps keep the process in perspective. Compression is simply a method for controlling and shaping dynamics in recorded audio. A good mix might not need any help in this department. Don't feel that you have to use a plug-in or processor just because you have it.

If you do choose to add compression, here are some practical tips for doing so:

- *To manipulate the "punch" in pop/rock mixes, start with a 2:1 ratio and a high threshold.* Lower the threshold until the compressor begins working, reducing a few dB on peaks. Set the release time to fast. Now, turn up the monitors a bit louder than normal, and adjust the attack time. At slow attack times, you should hear an accentuation of the accents in the music. Adjust the release time until the compressor grooves with the music. If the release is too fast, the compressor may start pumping—the sound of too much compressor action is a "fluttering" distortion. Too slow, and the compressor will not recover in time to allow the next peak through.

- *Watch your attack time.* Using a fast attack time lets less of the natural dynamic through the compressor before it acts to reduce the volume. A "zero" attack time (which is possible with modern plug-in compressors) will catch even the fastest of peaks. A fast attack with a high threshold can mildly reduce only the highest peaks in the audio, allowing you to increase the overall level. As you lower the threshold, the compressor will engage more often, smoothing out the dynamics in the music. This is the old "dig deep" technique, which increases a sense of average level and therefore loudness. Keep in mind that loudness is a trade-off with dynamics, therefore the "liveliness" of the music suffers the louder you get.

- *Bypass the compressor often, and adjust for volume differences when comparing.* Louder is not always better. Cutting too much range off the top of the dynamics and cranking the volume leads to a harsh, noisy impression of the music. The result: the "wimpy loud sound;" maximum level that sounds so harsh that listeners turn it down.

- *Multi-band compression is an excellent tool for modern music styles that contain strong bass/drum elements.* Typically, heavy bass sounds will trigger compression more readily than high frequencies, causing the whole mix to "pinch" in response to the low end. Multi-band compression allows you to adjust the compression for different frequency bands separately. You can create a particular punch in the low end with the technique described previously, while manipulating the mids and highs separately. You may want to use similar attack and release settings, to keep the compression consistent sounding, but use a lower ratio and higher threshold on high frequencies to prevent them from pumping with the low end. Some mastering engineers prefer to skip multi-band compression, and use a simpler compressor with a HPF (high-pass filter) on the audio, fed into the side chain input of the compressor. The HPFed side chain provides the compressor a bass-lite version of the audio to respond to, allowing the compressor to ignore heavy bass tones in the mix. This is another fine option for controlling the character of your compressor.

Again, talking about audio is like "dancing about football" as Billy Joel once put it. Refer to Chapters 7 and 12 on the CD-ROM for audio and visual information on compression. And please, treat the dynamics of your music with care.

Limiting in Mastering

Originally intended to catch only errant dynamics and keep them from causing overloads, the limiter has become the axis of discussion in the so-called "loudness war" happening in the audio industry.

Processors such as the T.C. Electronics' Finalizer and Waves' L1 and L2 Ultramaximizer plug-ins, although designed for transparent control of transient peaks in the best intentions of limiter design, have become the weapons of renegade engineers. Their target is our ears, and their goal is to make their recordings louder than everyone else's. It is essentially the modern version of Motown's loudness crusade, but this time it's getting unpleasant to listen to. Recent albums that have noticeable distortion from the abuse of limiting include the Red Hot Chili Peppers' "Dani California" and Metallica's "Death Magnetic." See Figure 12.1.

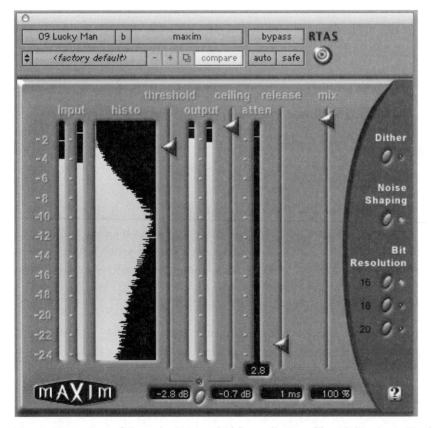

Figure 12.1   One of the culprits! Just kidding—limiters like DigiDesign's Maxim offer an excellent way to transparently reduce peaks in your audio and raise the overall level. The key to finding the right setting is to consider how much of your dynamics you want to trade for volume.

Considering that even good-quality consumer stereos cannot reproduce very fast transient peaks in the short-millisecond range, engineers can shave a few dB off these peaks without most listeners hearing the difference. With these few more dB of headroom freed up, you can boost the program material by that amount and create a louder recording. Only the audiophiles out there might hear the loss of these very fast transients if they could A/B some recordings. Oh well, what a pity for them; most people are listening on radios, boom boxes, and living room stereos, so who cares?

The question then became, "How much can we shave off the top of the recording before most people notice?"

The answer turned out to be, "always a little more than the last time!"

We can get used to just about anything, when it's fed to us in small increments. And that's just what happened. Over the last 10 years, producers and engineers have caved in to the pressure—from whoever it may come—to always make their recordings louder than the competition. And this is where the ultra-maximizers come in.

With most plug-in limiters, you can now simply dial in the limiter to reduce peaks by "x"dB, and it compensates by turning everything back up to just a hair below 0dBFS. This makes your recording loud. The more you lower the threshold, the louder it gets. These plug-ins are pretty transparent, so there's a tendency to think, "I wanna be loud like everyone else!"

And that's how it happens. dB by dB, we're crushing our music and undoing all those dynamics we spent all that time recording and mixing! You can do it, too. Wanna make a loud master? Just balance your stereo EQ, dial in some "dig deep" compression to raise the average level, and then limit the pants off the audio.

There's no reason to get into settings and tips—these limiter plug-ins are simple. The important skill is to know when enough is enough. In my opinion, this point should at least be *before* enough peaks in the audio get so flattened out as to sound like clipping (distortion)!

Musicians making recordings at home are dealing with many less-than-perfect circumstances. They need an edge to get noticed among the masses. Loud recordings get noticed, but do they have staying power? Sustained loudness is fatiguing, and listeners who don't know why they're irritated by your music eventually turn it off. So don't be self-defeating. Here are some realistic tips/arguments on limiting your audio:

- Let's say that some limiting is always necessary. If you've recorded acoustic music, especially with drums, there are bound to be a few stray peaks that need to be reeled in. Even if you want to maintain the smallest, fastest peaks in your music, you can set a limiter to keep an eye on unintentional peaks and to avoid overloads.

- I will go as far as to say that a couple dB can be shaved off the fastest peaks without anyone hearing this. As I noted, most playback systems won't reproduce these first few dB.

- Beyond this, you start making the trade. How important is it that your music be loud? Are you going to be playing back in bars, clubs, or other noisy spaces? In that case, average level becomes more important, and you probably should sacrifice some dynamics for loudness. Audiophile listeners and hypercritical mastering engineers aside, the typical listening environment does dictate how you should process music!

- If you are planning on pitching music to record labels, you have to deal with single-minded A&R types. (Oops, did I print that?) The catch 22 is that your CD must be loud, so that you'll grab their attention, but it shouldn't sound grating on the ears, or they'll just turn it off, and your CD becomes a coffee coaster. What to do? Consider making a fairly hot master of your "single"—put your best foot forward loudly. Then, find a more realistic balance between liveliness and "loudliness" for the rest of the songs. I think this is a fair suggestion. If they're interested after the first song, they will turn up the next tune—just make sure they don't want to turn it down.

- Experiment with limiting and the impression it makes when playing back your music on different playback systems. Try to balance dynamic and loudness so that there's still some life and excitement coming through on the smallest and lousiest of speakers. If you can manage that, your music will stand out in the crowd.

- If you insist that you like the sound of a crushed, loud recording, please consider processing a few individual tracks to sound this way; perhaps some electric guitars, synths, and drum loops. Limiting just the drum sub-group can open up a lot of dynamic space for other instruments. In any case, leave some dynamic, live, breathing sounds in your mix as a contrast! This is the best of both worlds, and makes for an attention-grabbing and interesting mix. Laying this on the whole mix is just lazy.

Dithering

Dithering is the term for the mathematical process used to reduce the bit rate of your audio. CDs still use the 16 bit/44.1kHz sample rate standard, so if you have recorded at 24 bits, you will need to reduce these 24 bits to 16.

It is important not to simply output a 24-bit signal into a 16-bit file. This is possible, but results in a truncated (cut off) signal. If you simply lop off eight bits of the signal, you'll lose a lot of subtle stereo imaging and ambience.

To avoid chopping off those bits, dithering adds a random eight-bit number to the least significant eight bits of each 24-bit sample, and rounds the result to the lowest bit of a resulting 16-bit number. This is done for every sample! It's a lot of math being done to your signal, but that's happening to digital audio all the time.

This random 8-bit number (the dither) actually sounds like noise—hiss. It is very low level (down at −96dBFS), but significantly helps to maintain subtle ambience and stereo imaging. Clever programmers have discovered that it is possible to use noise that has a different frequency content than just random hiss (white noise). This is called "noise-shaped dither," and reduces the audibility of the noise added during the conversion from 24 to 16 bits.

Many DAWs have a few dithering programs available, and you can test your dither to determine what you prefer. Simply record a bit of room tone (silence) at 24 bits. Insert dither on this track, and crank up the volume. Switch the dither on and off, and try different dithers. Pick the least obtrusive, least noisy, or least distorted sounding option.

Last Thoughts on Mastering

After making all your meticulous adjustments, bypass all those wonderful processors and just listen to the plain old mix again. Did you make it worse? Or just dress up a pig? Be honest!

I often find that I've EQed and tweaked far too much, and that the mix needs less adjusting than I initially thought. This goes for compression most often, and I wind up reducing the ratio and increasing the threshold.

Be sure to compare processed and unprocessed (bypassed effects) audio at the same volume. As noted—over and over again—compression and limiting often sound "good" just because they sound loud. Hearing the original dynamics come through at the same volume may be enough to convince you to back off on the compression/limiting. Refer to Chapter 12 on the CD-ROM for some audio to back up this argument!

The luxury of the home studio is flexibility. Just because you are in the mastering stage of the project doesn't mean that you can't go back and tweak a mix. If you find yourself trying to correct mix problems with EQ, please just go back and fix the mix, run a new file, and then return to mastering. Take advantage of this flexibility to make your music sound its best—carving up the mix with EQ during mastering is not going to do this.

Keep the big picture in mind at all times—every song has an overall dynamic shape; like a roller coaster; it has quiet passages and high points. Make sure that the overall shape excites the listeners and conveys the emotional message that the artist intended. This is key to creating a good master recording.

A sense of dynamics should continue throughout the record as well. Consider which song should open the album, and which wraps it up best. Where should the "hit single" appear? Consider transitions—how much space between songs and how songs sound next to one another. You may want to avoid a song in E and one in B♭ right next to each other (that ♭5 interval may make the following song sound dissonant until the ear adjusts). If it's a heavy metal album, this may be just what you want!

Be especially sure that the relative levels of songs make sense. A ballad (which may have fewer transients, and can therefore be made louder quite easily) shouldn't wind up louder than the up-tempo rock track that follows!

MP3 formatting comes next—you are making MP3s, right? You must! Once your masters are done, create copies in yet another folder, and set aside some time for MP3 conversions. Try different encoders and do some critical listening. Not all MP3 encoding programs are created equal.

Finally, as a mastering engineer, do your homework about the duplication process. You'll probably be making CDs, but some bands (such as punk and electro bands) still make vinyl singles. Find out what the printers need, and be sure you deliver them in the proper format. You might need a CD authoring program that provides a particular type of encoding, so ask ahead.

Index

M

machine isolation pads, 61
Martin, George, 14–16
Master Handbook of Acoustics, 59
mastering
   advertisements, 239
   defined, 28–29
   formats, 238
   gear, 239–240
   mixing, 238
   overview, 237–238
   studios, 238
   techniques
      compression, 245–246
      dithering, 249–250
      EQ, 243–245
      limiters, 247–249
      overview, 250–251
      preparation, 239–242
      processing chain, 242–243
      stereo wideners, 245
   volume, 238–239
matresses (studios), 59
meters (limiters), 158–159
microphones
   acoustic guitars, 98–99
   bass guitars
      choosing, 102–103
      DI boxes, 102–103
      distortion, 104–105
      setup, 104
      techniques, 104–105
      troubleshooting, 105–106
   brass, 108–109
   buying, 37–39
   characteristics, 40–41
   choosing, 37–39, 52–53
      acoustic guitars, 98
      bass guitars, 102–103
      brass, 108
      drums, 112–114
      electric guitar, 99–101
      keyboards, 106
      organs, 106
      percussion, 110
      pianos, 95
      stringed instruments, 109
      unusual instruments, 125
      upright basses, 102–103
      vocals, 91
   close mic'ing, 6, 10
   condenser microphones, 36–37, 39
   cost, 37–39
   DI boxes, 50, 102–103

diaphragm microphones, 31–35
diaphragm size, 44
drums
   choosing, 112–114
   drum machines, 123
   electronic drums, 123
   loops, 124
   MIDI drums, 123–124
   multiple microphones, 117–119
   overview, 111
   samples, 124–125
   sequencers, 123
   setup, 114–119
   techniques, 120–121
   three microphones, 116–117
   troubleshooting, 121–125
   two microphones, 115–116
dynamic microphones, 35–36
electric guitars
   choosing, 99–101
   techniques, 101–102
   troubleshooting, 102
EQ, 41
FET, 50–52
frequency response, 41–44
interfaces, 85–86
keyboards, 106–107
mixing techniques, 226–227
multiple. See multiple microphones
organs, 106–107
overview, 31–32
percussion, 110–111
phase
   in phase, 54
   out of phase, 54
   phase shift, 54–55
   polarity switch, 55–56
   problems, 44
   relativity, 53–54
   three-to-one rule, 54–55
pianos, 95–97
piezoelectric crystal microphones
   defined, 32
   overview, 39–40
polar patterns, 45–49
   bleed, 46
   cardioid, 45–49
   figure-eight, 48
   hyper-cardioid, 46–49
   omni-directional, 47–48
   selectable, 49
preamps, 85–86
ribbon microphones
   defined, 31
   frequency response, 32–33

COURSE TECHNOLOGY
CENGAGE Learning
Professional • Technical • Reference

Course Technology PTR
COURSE CLIPS

Introducing *Course Clips*!

Course Clips are interactive DVD-ROM training products for those who prefer learning on the computer as opposed to learning through a book. *Course Clips Starters* are for beginners and *Course Clips Masters* are for more advanced users.

Pro Tools 8
Course Clips Master
Steve Wall ■ $49.99

Pro Tools 8
Course Clips Starter
Steve Wall ■ $29.99

Ableton Live 8
Course Clips Master
Brian Jackson ■ $49.99

Individual movie clips are available for purchase online at **www.courseclips.com**

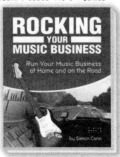

License Agreement/Notice of Limited Warranty

By opening the sealed disc container in this book, you agree to the following terms and conditions. If, upon reading the following license agreement and notice of limited warranty, you cannot agree to the terms and conditions set forth, return the unused book with unopened disc to the place where you purchased it for a refund.

License:
The enclosed software is copyrighted by the copyright holder(s) indicated on the software disc. You are licensed to copy the software onto a single computer for use by a single user and to a backup disc. You may not reproduce, make copies, or distribute copies or rent or lease the software in whole or in part, except with written permission of the copyright holder(s). You may transfer the enclosed disc only together with this license, and only if you destroy all other copies of the software and the transferee agrees to the terms of the license. You may not decompile, reverse assemble, or reverse engineer the software.

Notice of Limited Warranty:
The enclosed disc is warranted by Course Technology to be free of physical defects in materials and workmanship for a period of sixty (60) days from end user's purchase of the book/disc combination. During the sixty-day term of the limited warranty, Course Technology will provide a replacement disc upon the return of a defective disc.

Limited Liability:
THE SOLE REMEDY FOR BREACH OF THIS LIMITED WARRANTY SHALL CONSIST ENTIRELY OF REPLACEMENT OF THE DEFECTIVE DISC. IN NO EVENT SHALL COURSE TECHNOLOGY OR THE AUTHOR BE LIABLE FOR ANY OTHER DAMAGES, INCLUDING LOSS OR CORRUPTION OF DATA, CHANGES IN THE FUNCTIONAL CHARACTERISTICS OF THE HARDWARE OR OPERATING SYSTEM, DELETERIOUS INTERACTION WITH OTHER SOFTWARE, OR ANY OTHER SPECIAL, INCIDENTAL, OR CONSEQUENTIAL DAMAGES THAT MAY ARISE, EVEN IF COURSE TECHNOLOGY AND/OR THE AUTHOR HAS PREVIOUSLY BEEN NOTIFIED THAT THE POSSIBILITY OF SUCH DAMAGES EXISTS.

Disclaimer of Warranties:
COURSE TECHNOLOGY AND THE AUTHOR SPECIFICALLY DISCLAIM ANY AND ALL OTHER WARRANTIES, EITHER EXPRESS OR IMPLIED, INCLUDING WARRANTIES OF MERCHANTABILITY, SUITABILITY TO A PARTICULAR TASK OR PURPOSE, OR FREEDOM FROM ERRORS. SOME STATES DO NOT ALLOW FOR EXCLUSION OF IMPLIED WARRANTIES OR LIMITATION OF INCIDENTAL OR CONSEQUENTIAL DAMAGES, SO THESE LIMITATIONS MIGHT NOT APPLY TO YOU.

Other:
This Agreement is governed by the laws of the State of Massachusetts without regard to choice of law principles. The United Convention of Contracts for the International Sale of Goods is specifically disclaimed. This Agreement constitutes the entire agreement between you and Course Technology regarding use of the software.